Fit for hot and cold rolling of strips

Exercises

Fit for hot and cold rolling of strips

Exercises

by

Michael Degner and Heinz Palkowski

Photos: SMS group GmbH

Bibliografische Information der Deutschen Nationalbibliothek:
Die Deutsche Nationalbibliothek verzeichnet diese Publikation in der Deutschen Nationalbibliografie; detaillierte bibliografische Daten sind im Internet über http://dnb.dnb.de abrufbar.

© 2016 Michael Degner and Heinz Palkowski

Illustration: Paul Westerwalbesloh
Photos: SMS group GmbH

Herstellung und Verlag: IMP InterMediaPartners GmbH, Wuppertal

ISBN: 978-3-9817904-2-9

– Contents –

Preface 6
0. Abbreviations 7
1. Mass flow law in HSM's and CRM's 8
2. Roll force and roll torque calculation in CRM's 19
3. Design and Dimensioning of HSM's 22
4. Use of a CB in HSM's 39
5. Process automation in HSM's 47
6. Thickness control in HSM's 69
7. Measuring techniques in HSM's 83
8. Roll flattening and minimum rollable strip thickness ... 105
9. Operational data of HSM's 110
10. Commissioning of rolling mills 117
11. Materials for rolls and wear of rolls 122
12. Questionnaire of rolling mill technique 125
13. Bibliography 132
14. List of key words 137

Preface

This compilation of exercises with further explanations of technological aspects in the field of hot and cold rolling of strip material is the prosecution of the books "Mathematics for physicists and engineers – Basics" and "Technology of Hot Strip Production – Calculation examples and exercises" and "Modern Hot Strip Production – Process and Plant Technology". The latter one gives the theoretical and technical information for process understanding and the basic information for solving the exercises.

This book is focusing on engineers at site in Hot Strip Mills and Cold Rolling Mills and students of rolling techniques. The addressed and discussed problems are near to practice and give students and engineer's hints for answering typical daily questions. For metal forming students this book guarantees assistance due to preparation of examinations at the universities. Moreover, it helps to understand the process and its side effects. According to the slogan "From a practitioner to practitioners" a complete derivation of the rolling theory is not given in the formulation of the necessary basic rolling formula. Theory is introduced only as far as necessary to get a meaningful understanding of the hot and cold rolling process. The complete theory for hot and cold rolling is given in the relevant basic literature of metal forming technology and in the book "Basics in Flat Rolling" compiled by the authors of this exercise book.

We wish fun and joy calculating and studying the exercises with their additional explanations and supporting comments.

Michael Degner and Heinz Palkowski

0. Abbreviations

In this book the following abbreviations are used:

BUR	Back-Up Roll
CB	CoilBox
CS	Crop Shear
CRM	Cold Rolling Mill
DC	Down Coiler
FM	Finishing Mill
FS	Finishing Stand
HPM	Hot Plate Mill
HSM	Hot Strip Mill
MRS	Multi Roller Stand
RM	Roughing Mill
RS	Roughing Stand
WR	Work Roll

B, b	[mm]	width
H, h	[mm]	thickness
L, l	[mm, m]	length
m	[kg, t]	mass
T	[K,°C]	temperature
t	[s, min, hour, a]	time
v	[m s^{-1}, m min^{-1}]	speed
ρ	[t m^{-3}]	density

Indices
0	entry/start
1	exit/end
e	exit

1. Mass flow law in HSM's and CRM's

The mass flow is one of the most important values to be controlled in a rolling mill. The basis for this value is the constancy of volume for forming processes. Therewith the geometrical changes can be calculated for each forming step as well as dependencies, such as changes in speed /1/.

Exercise 1-1: Final rolling thickness and rolling speed

A transfer bar with a thickness of $h_0 = 44$ mm is fed into the FM of a HSM with a speed of $v_0 = 0.5$ m s^{-1}. Calculate the final strip thickness h_1 assuming that the rolling speed is $v_1 = 11$ m s^{-1} for the strip head end.

Solution:

Regarding the law of volume constancy

$$l_0 \cdot h_0 \cdot b_0 = l_1 \cdot h_1 \cdot b_1 \tag{1.1.1}$$

for rolling without spread ($b_0 = b_1$) in the FM the following equation can be derived:

$$h_0 \cdot v_0 = h_1 \cdot v_1. \tag{1.1.2}$$

Converting the equation and inserting the given values the final strip thickness follows according to:

$$h_1 = \frac{h_0 \cdot v_0}{v_1} = \frac{44 \text{ mm} \cdot 0.5 \text{ m s}^{-1}}{11 \text{ m s}^{-1}} = 2 \text{ mm}. \qquad (1.1.3)$$

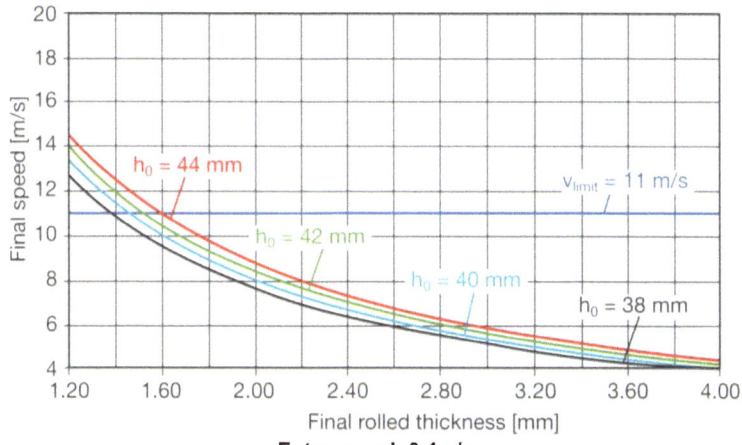

Entry speed: 0.4m/s

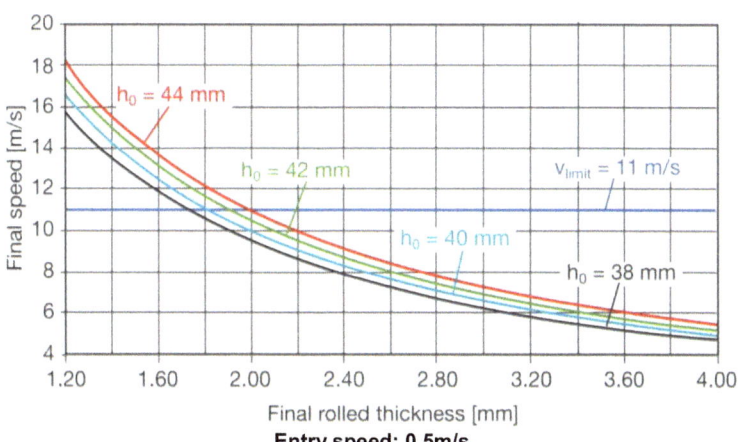

Entry speed: 0.5m/s

Fig. 1.1a: Relation between final rolled thickness and final speed for different strip entry speeds into the FM (Parameter: transfer bar thickness)

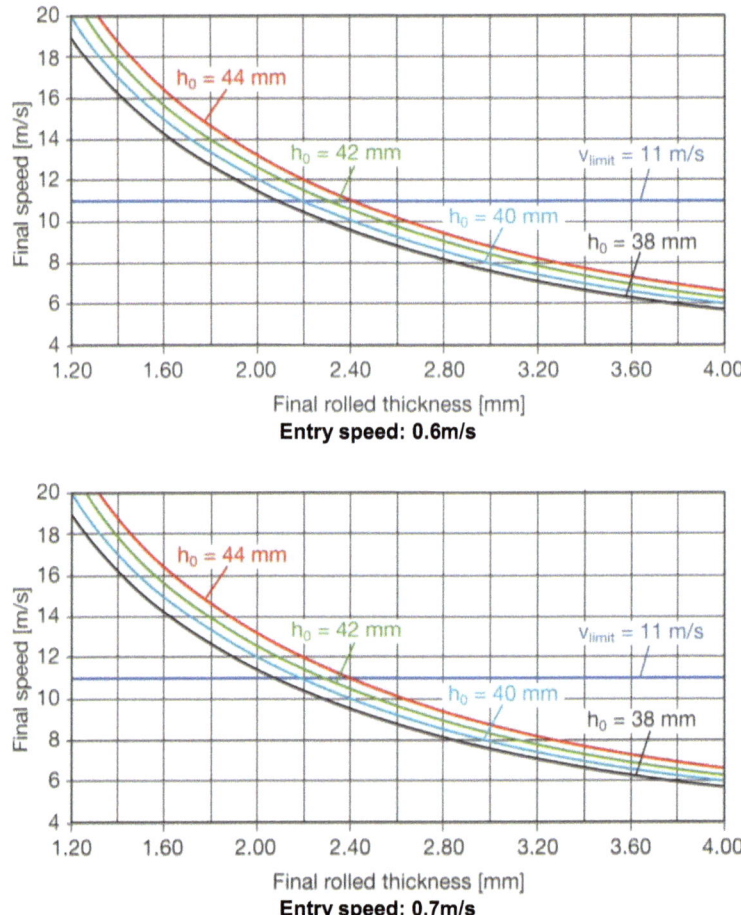

Fig. 1.1b: Relation between final rolled thickness and final speed for different strip entry speeds into the FM (Parameter: transfer bar thickness)

Fig. 1.1a and **1.1b** illustrate the relationship between final rolling speed and final strip thickness for different bar thicknesses and entry speeds into the FM. The given maximum strip head end speed (11 m s^{-1}) is plotted in the graphs, too. Increasing the transfer bar thickness and/or entry speed shifts the possible final rolled strip thickness to higher values.

Remark

Rolling without spread assumes that the ratio between rolling width b and rolling thickness h is larger than 10. In the following these conditions are discussed for HSM's. In conventional slab casting lines the cast slab thickness normally is in between 350 mm and 250 mm, for thin slab casting the cast thickness is in between 50 mm and 90 mm. The cast width can amount up to 2,200 mm; in HSM's slab widths are between 800 mm and 2,200 mm. In medium strip width mills slabs between 200 mm and 800 mm and in small width strip mills rectangular billets with sizes as e.g. 200 mm x 200 mm are rolled. For hot strip production the slabs are reduced to transfer bars in a RM in three to seven passes, depending on the slab thickness, quality of the steel and the finishing conditions. **Table 1.1** gives the ratio for rolled material width to thickness for different plant types and geometries.

Parameter	Hot wide strip	Medium strip	Small strip	Thin slab
Slab/Billet width [mm]	1,000	400	150	1,000
Slab/Billet thickness [mm]	250	250	150	80
Transfer bar width [mm]	1,000	400	100	1,000
Transfer bar thickness [mm]	40	40	40	40
Final strip width [mm]	1,000	400	100	1,000
Final strip thickness [mm]	2	2	2	2
Ratio b/h slab/Billet [-]	4	1.6	0,66	12,5
Ratio b/h Transfer bar [-]	25	10	2,5	25
Ratio b/h Final strip [-]	500	200	50	500

Table 1.1: Spreading behavior during rolling of flat products. b/h-ratio for several HSM configurations

This means, material spreading has to be considered in pass schedule calculations in all treated cases in the area of the RM and even for medium and small strip rolling mills in the FM.

Exercise 1-2: Casting speed and casting thickness

In a two strand casting rolling plant (n=2) an amount of m = 300 t steel per hour is to be cast. The cast width is fixed at b=1,300 mm. Casting speed is set to $v = 5$ m min^{-1} and steel density is given with $\rho = 7.5516$ t m^{-3}. Calculate the thickness of the cast thin slab.

Solution:

The cast thickness in the strands is calculated using the equation of mass flow according to

$$h = \frac{m}{\rho \cdot t \cdot n \cdot b \cdot v}$$

$$h = \frac{300 \text{ t}}{7.5516 \text{ t m}^{-3} \cdot 60 \text{ min} \cdot 2 \cdot 1.3 \text{ m} \cdot 5 \text{ m min}^{-1}}. \qquad (1.2.1)$$

$h = 51$ mm

The cast thickness is $h = 51$ mm.

Fig.1.2 shows the resulting cast thicknesses for different cast widths and speeds. Depending on quality and solidification conditions casting speeds up to 7 m/min actually are used in thin slab casting. The thickness of thin slabs is between 50 mm and 90 mm. Slabs with a thickness up to 350 mm and a width up to 2,200 mm are produced on conventional thick slab casting plants and used as feeding stock in conventional high capacity HSM's. One example for a high capacity HSM is the 3/4-continuously operating ThyssenKrupp HSM at Duisburg-Beeckerwerth with a capacity of approximately 6 million tons a year. Materials with a cast thickness of 400 mm are produced for further rolling in HPM's. One example for a high capacity HPM is the rolling mill of the Dillinger Hütte at Dillingen site. The yearly capacity of HPM's with one RS and one FS is about 1.5 million tons.

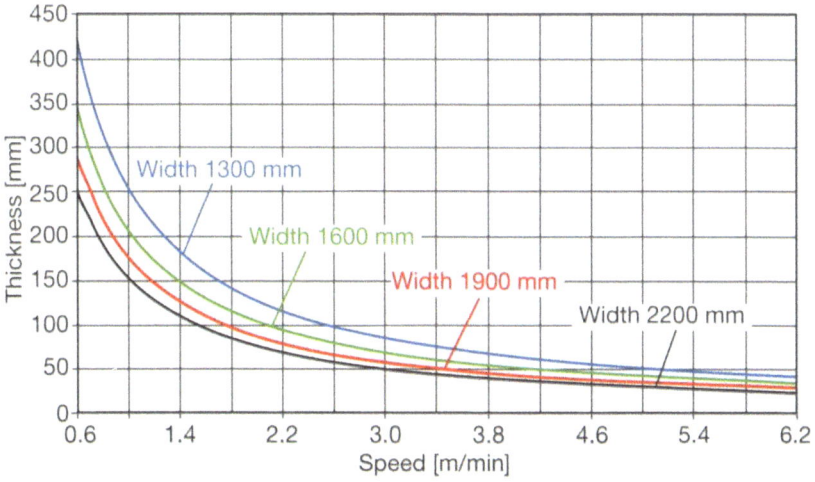

Fig.1.2: Casting thickness for different casting speeds

Exercise 1-3: Production of a HSM

Besides its layout, the capacity of a mill strongly depends on the program to be rolled. As thinner the final strip the longer the running time in the mill will be and therewith the productivity decreases; the strip width influences the productivity linearly.

A 3/4-continuously operating HSM produces m = 5 million tons hot strip per year. The average final strip thickness is h = 3.55 mm and the average final width b = 1,340 mm. Imagine, the total finished rolled strip is welded together. How often can the resulting total strip length be wound around the earth?

The average hot strip density is assumed to be $\rho = 7.86 \text{ t m}^{-3}$.

Solution:

By transposing the relationship $\rho = \dfrac{m}{V} = \dfrac{m}{b \cdot h \cdot l}$ the average length l of the endless strip is

$$l = \frac{m}{\rho \cdot b \cdot h}. \qquad (1.3.1)$$

Using the given values the endless strip length is calculated according to:

$$I = \frac{m}{\rho \cdot b \cdot h} = \frac{5 \cdot 10^6 \, t}{7.86 \, t \cdot m^{-3} \cdot 1.40 \, m \cdot 0.00355 \, m} . \qquad (1.3.2)$$

$I = 133,726 \, km$

The circumference of the earth is 40,000 km roughly. The yearly strip production of the HSM can be wound approximately 3.4-times around the earth.

Exercise 1-4: Rolling of strip in a CR-tandem mill

A hot wide strip with a thickness of 1.8 mm is rolled down in a five-stand tandem mill to a final thickness $h_1 = 0.40 \, mm$. The thickness for the single passes is given in **Table 1.2**.

Stand	1	2	3	4	5
Exit thickness [mm]	1.34	0.89	0.62	0.45	0.40

Table 1.2: Strip thickness distribution in a CRM

The final rolling speed is 1,500 m min^{-1}. Calculate the strip entry speed into the mill and the strip speeds between the stands in m s^{-1}.

Solution:

First, the final rolling speed has to be shifted into the unit [m s^{-1}]:

$$v_e = 1,500 \, \frac{m}{min} \cdot \frac{min}{60 \, s} = 25 \, m \, s^{-1} \qquad (1.4.1)$$

The entry speed in the tandem mill can be calculated as follows:

$$v_0 = \frac{h_1}{h_0} \cdot v_1 = \frac{0.4 \text{ mm}}{1.8 \text{ mm}} \cdot 25 \, \frac{m}{s} = 5.55 \, \frac{m}{s}. \qquad (1.4.2)$$

Therewith, the strip exit speed for each mill stand follows:

$$v_i = \frac{h_{i+1}}{h_i} \cdot v_{i+1}. \qquad (1.4.3)$$

The results for the strip exit speeds are summarized in **Table 1.3**.

Stand number i	1	2	3	4	5
Thickness ratio $\frac{h_{i+1}}{h_i}$ [-]	0.299	0.449	0.645	0.889	1.0
Exit speed [m s^{-1}]	7.48	11.23	16.13	22.23	25.00
Exit speed [m min^{-1}]	449	674	968	1,334	1,500

Table 1.3: Strip speed distribution in a CRM

Exercise 1-5: Spreading in HSM roughing stands

A slab with the dimensions $h_0 = 200$ mm and $b_0 = 1000$ mm is rolled in a semi-continuous HSM. The thickness draft at the first pass in the RM is $\varepsilon_1 = 23.3\%$ and the roll diameter of the first RS is given with $d = 1200$ mm.

Calculate the slab width spreading of the first pass in the RS using the models according to:

a.) Geuze
b.) Tafel und Sedlaczek
c.) Siebel
d.) Hill
e.) Wusatowski

Solution:

For calculation of slab spreading the knowledge of thickness draft and slab exit thickness is necessary. These parameters can be calculated according to:

$$\varepsilon_h = \frac{h_0 - h_1}{h_0} \tag{1.6.1}$$

and

$$\Delta h = h_0 - h_1 \tag{1.6.2}$$

Inserting the given parameters provides

$\Delta h = \varepsilon \cdot h_0$
$\Delta h = 0.233 \cdot 200$ mm
$\Delta h = 46.6$ mm

and therewith $h_1 = 153.4$ mm.

The relevant spreading equations are given in Chapter 6 of the book "Basics in Flat Rolling" /27/. The solutions for strip spreading according to the different models are listed in **Table 1.4**.

Model	Geuze	Tafel	Siebel	Hill	Wusatowski
Spreading [mm]	16	28	14	7	1
Spreading [%]	1.6	2.8	1.4	0.7	0.1

Table 1.4: Spreading calculation in a HSM (First pass RS)

Therefore strip spreading in HSM RS's should be considered in calculations. As can be seen, depending on the assumptions made and parameters used the result differs strongly. So, decision has to be made which boundary conditions are given for the mill used to select a proper equation!

2. Roll force and roll torque calculation in CRM's

The equations for calculation of roll force and roll torque in CRM's are summarized in Chapter 14 of the book "Basics in Flat Rolling" /27/. For completeness of the calculations in this exercise book the relevant equations are listed here once again.

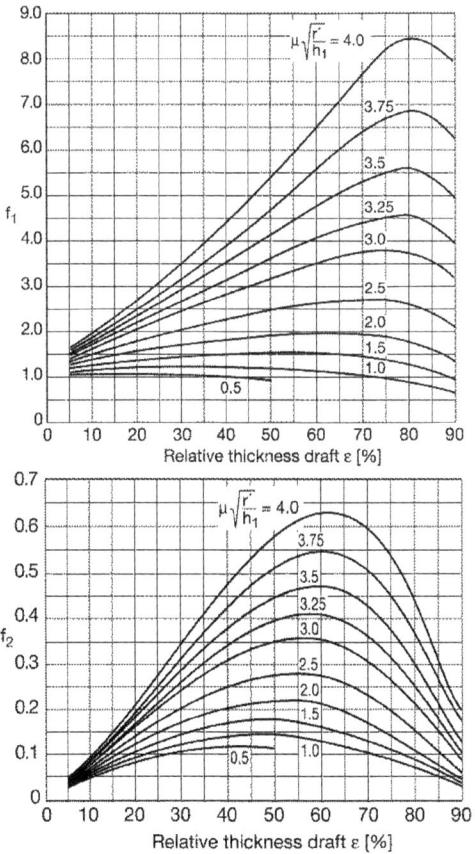

Fig. 2.1: f_1- and f_2-function for determining roll forces and roll torques in cold rolling without applied strip tensions according to /9/

The roll force in cold rolling without applied tensions is:

$$F = k_{fm} \cdot b \cdot \sqrt{r' \cdot (h_0 - h_1)} \cdot f_1\left(\mu \cdot \sqrt{\frac{r'}{h_1}}, \varepsilon\right). \tag{2.1}$$

The roll torque is calculated according to:

$$M = 2 \cdot r \cdot \frac{h_0^2}{h_1} \cdot k_{fm} \cdot b \cdot f_2\left(\mu \cdot \sqrt{\frac{r'}{h_1}}, \varepsilon\right). \tag{2.2}$$

k_{fm} represents the average yield strength, b the strip width, r' the flattened roll radius, h_0 and h_1 the entry and exit thickness, respectively, ε the relative thickness reduction and r the non-flattened work roll radius.

The functions f_1 and f_2 describe the influence of the roll gap. They are illustrated in **Fig. 2.1**.

Exercise 2-1: Roll force and roll torque in cold rolling without strip tension

In a CRM the following strip is rolled: Steel grade S235JR, work roll radius $r = 200$ mm, entry thickness $h_0 = 3.0$ mm, strip width $b = 1200$ mm, average yield stress $k_{fm} = 454$ MPa, friction coefficient $\mu = 0.1$ and absolute thickness draft $\Delta h = h_0 - h_1 = 0.9$ mm. Calculate the roll force and roll torque according to Ford and Bland. Interpolate the resulting values for the functions f_1 and f_2 from the nomograms in Fig. 2.1.

Solution:

The roll force using Ford and Bland algorithm without applying strip tensions at the entry and exit of the stand is given by equation (2.1) and with $f_1(0.975, 0.3)=1.2$ interpolating from the nomograms in Fig. 2.1:

$$F = 454\ \frac{N}{mm^2} \cdot 1200\ mm \cdot \sqrt{200\ mm \cdot (3-2.1)\ mm} \cdot 1.2 \quad (2.3)$$

$F \approx 8.8\ MN$

The resulting roll torque is calculated using equation (2.2) and $f_2(0.975, 0.3) = 0.12$ interpolated from the nomograms in Fig. 2.1 according to:

$$M = 2 \cdot 200\ mm \cdot \frac{3^2\ mm^2}{2.1\ mm} \cdot 454\ \frac{N}{mm^2} \cdot 1200\ mm \cdot 0.12 \quad (2.4)$$

$M \approx 112\ kNm$

3. Design and Dimensioning of HSM's

Tables 3.1 and **3.2** give the facilities, their functions and configuration in the different areas of a HSM in direction of material flow and typical technical data of a high capacity 3/4-continuously operating HSM.

Area	Facility	Remark, function	Necessity
Slab reheating	Slab yard	Slab supply	always
	Reheating furnace	Slab reheating up to 1,250°C	always
Roughing train	Descaler	Primary scale removal with pressures between 150 and 200 bar	always
	Slab sizing press	Width reduction up to 350 mm	optional
	2-high roughing stand	1 or 3 horizontal passes	optional
	4-high-reversing roughing stand	5 to 7 (9) horizontal passes	always
	Edger	Width reduction and width control	always
	Coilbox	Material- and heat buffer	optional
	Heat cover shield	Heat buffer	optional
Finishing train	Crop shear	Cropping transfer bar head and tail	always
	Inductive edge heater	Heating strip edges up to 100 K	optional
	Descaler	Secondary scale removal with pressures between 200 and 400 bar	always
	Finishing stands	Horizontal passes, 5 to 7 stands	always
	Looper	Tension- and loop control	always
	Tensiometer-Looper	Tension measurement over strip width	optional
Exit area	Run out table	Strip transport to down coilers	always
	Laminar cooling	Adjusting of mechanical properties	always
	Edge masking	Reduction of wavy edges	optional
	Compact cooling	Extended cooling strategies	optional
Down coiler	Universal down coiler	3-roller coiler, 1 to 3	always
	4-roller down coiler	For thick gage strip	optional
	Coil transport system	Coil transport, inspection, weighing and marking	always
Automation	Level 0	Drive control, sensoring	always
	Level 1	Technological control	always
	Level 2	Technological process models	always
	Level 3	Product data information (PDI-data), customer	always
Measuring	Geometry	Width, thickness, profile and contour, flatness	always
	Temperature	Strip center position	always
		Scan over width	optional
	Surface condition	Top side behind finishing train	optional
		Bottom side before down coiler	optional
	Coil telescopicity	Coil transport system	optional

Table 3.1: Facilities in a HSM according to Spur /34/

	Type of mill stand	Roll diameter		Mill stand power	Mill stand motor revolution	Rolling speed	Gear ratio	Torque	
		WR [mm]	BUR [mm]	[kW]	[rpm]	[m/s]		Motor [kNm]	Rolls [kNm]
R1	2-high	1,350		6,000	375	1.47	18,00	153	2,752
R2	4-high reversing	1,200	1,600	2 x 9,000	50/100	3.14/6.28		3,440/ 1,720	3,440/ 1,720
F1	4-high reversing	900	1,600	11,000	165/300	1.41/2.57	5,50	637/ 350	3,503/ 1,927
F2	4-high	900	1,600	11,000	165/400	1.73/4.19	4,50	637/ 263	2,866/ 1,182
F3	4-high	900	1,600	11,000	165/400	2.59/6.28	3,00	637/ 263	1,911/ 788
F4	4-high	750	1,600	11,000	165/400	4.05/9.82	1,60	637/ 263	1,019/ 420
F5	4-high	750	1,600	11,000	165/400	6.48/15.71	1,00	637/ 263	637/ 263
F6	4-high	750	1,600	8,500	165/400	7.62/18.46	0,85	492/ 203	418/ 173
F7	4-high	750	1,600	8,500	165/400	9.26/22.44	0,70	492/ 203	345/ 142

Table 3.2: Technical data of a conventional high capacity hot strip mill according to Spur /34/

The data given in Table 3.2 represent a HSM able to roll strip widths up to 2,030 mm and final strip thicknesses in between 1.5 mm and 25 mm; typical pass schedules for different steel grades are given in **Table 3.3**. The treated steel grades thereby are:

- Soft unalloyed construction steel grade S235JR, final dimension 25 mm x 2,030 mm,
- Austenitic stainless steel grade X 5 CrNi 1810, final dimension 4.5 mm x 1,830 mm,
- Micro alloyed pipe steel grade X70, final dimension 16 mm x 2,030 mm.

Remarkable are the different final rolling temperatures at the exit of the finishing train, the number of passes in the

RM (5 to 7 passes), the transfer bar thickness between 40 mm and 50 mm and, finally, rolling the steel grade X70 with five stands in the FM.

Steel grade	Charging temp. [°C]	Slab thickness [mm]	Roughing train							Strip head temperature [°C]	
			Thickness rolling pass [mm]							after last pass	before shear
			Relative draft [%]								
S235JR	1,250	250	220	166	115	84	60	45		1,119	1,084
			12	25	31	27	29	25			
X 5 Cr Ni 18 10	1,350	250	220	167	121	85	58	40		1,149	1,105
			12	24	28	30	32	31			
X70	1,250	250	230	195	162	131	104	81	63	50	1,061	950
			8	15	17	19	21	22	22	21		

Steel grade	Temp. before shear [°C]	Transfer bar thickness [mm]	Finishing train							Speed last pass (Head)	Temp. F7 (Head)
			Thickness rolling pass [mm]								
			Relative draft [%]								
			F1	F2	F3	F4	F5	F6	F7		
S235JR	1,084	45	25.20	1.45	7.03	4.61	3.42	2.84	2.50	8.43	880
			44	51	44	34	26	17	12		
X5Cr-Ni 18 10	1,105	40	26.78	16.77	11.33	8.13	6.27	5.23	4.5	6.86	1,000
			33	37	32	28	23	17	14		
X70	950	50	3.90	29.16	22.66	18.57	16.00			1.41	800
			24	23	22	18	14				

Table 3.3: Pass schedule roughing and finishing train for different steel grades /34/

Exercise 3-1: Bottle neck consideration in HSM's

A semi-continuously operating HSM may be designed with a slab reheating furnace, one reversing RS, a CB and six FS's. Part of the production are pipe steel grades according to **Table 3.4**. Complete the missing pass schedule data! Take into account that theoretical and practical strip cycle times may differ due to an overloading and following necessary cooling down of the main drives.

Piece number	1	2	3
Rolled steel grade	X70	X70	X70
Furnace capacity [t/h]	450	450	450
Rolling time RM[s]	140	196	162
Theoretical cycle time RM[s]	185	241	207
Cycle time RM in reality [s]	185	241	207
Rolling time FM [s]	156	111	107
Theoretical cycle time FM[s]	201	156	152
Cycle time FM in reality [s]	264	181	192
Final thickness [mm]	6.00	14.00	14.00
Final width [mm]	1650	1650	1650
Transfer bar head end temperature [°C]	1012	954	1005
Transfer bar tail end temperature [°C]	1038	969	908
Coil weight [t]	29.70	29.70	29.70
Rolling with CB (yes/no)			
Production RM[t/h]			
Production FM [t/h]			
Bottle neck (Furnace/RM/FM)			

Table 3.4: Pipe steel grade X70-Pass schedule data

Solution:

Piece number	1	2	3
Rolling with CB (yes/no)	yes	yes	no
Production RM [t/h]	578	444	517
Production FM [t/h]	405	591	557
Bottle neck (Furnace/RM/FM)	FM	RM	Furnace

Table 3.5: Pipe steel grade X70-pass schedule missing data

The production P [t h^{-1}] in the RM and FM follows - together with the cycle time t [s] and the coil weight m [t]:

$$P = \frac{3{,}600}{t} \cdot m. \qquad (3.1.1)$$

Rolling with and without CB can be derived from the temperatures of the transfer bar. When using CB operation mode the transfer bar temperatures at the transfer bar head are minor than at its tail end. The bottle neck is given by comparison of production figures of furnace, RM and FM. It is the minimum of the three production values.

Remark

This kind of investigation forms the basis for mill-pacing, i.e. the configuration of rolling schedules under the aspect of maximizing the mill productivity taking into account mill design and boundary conditions. Due to probably necessary cooling down times of the main drives in the FM- because of overload - the strip cycle times are prolonged in the above exercise and the FM becomes the bottle neck, **Table 3.5**. These considerations are important for dimensioning and the layout of the rolling mill because a special product mix with their corresponding steel grades and final geometries needs to be produced. The total costs for mill design and its later operating are influenced mainly from this kind of considerations. This is the main task of a plant builder: Configuration of a suitable plant concept for fulfilling the requested product mix and productivity combined with an adapted cost structure.

In trouble free plant condition mill pacing automation provides optimizing mill productivity by discharging slabs from the reheating furnace at correct times. This can be managed by e.g. traffic light signals to the operating personnel or even in automatic mode. Time loss for production can be prevented and mill productivity and yield can be increased overall.

Furthermore, mill-pacing investigations allow the derivation of nominal cycle times for the strip rolled and therefore for total rolling sequences within the rolling schedule. The comparison of nominal with actual measured cycle times is possible. Deviations of actual and theoretical mill-pacing times can be caused by:

- Essential views and checks of the finished rolled product by naked eye of the operating personnel,

- Necessary checks at the mill in advance before rolling critical grades and dimensions.

The comparison of nominal and actual cycle times leads to the definition of further numbers in the time-productivity scheme of rolling plants, explained in the following.

General considerations on operation numbers in production facilities

In some production plants numbers like utilization time, operation time and run-time tables are introduced for an objective assessment of the mill or even the complete plant.

The question arises why this kind of assessment is more objective than in former times? An example for skin-passing in a CRM facility is taken for illustration.

A retrospect:

Production shift A of a skin pass mill has to roll O5-strip material for exposed automotive applications during the complete shift; the shift productivity is lowered because of the high quality demanded. If the planning department randomly scheduled further O5-rolling campaigns for shift A several times in a month, productivity decreases further. Latest at the end of the year no one remembers that the supposed lower shift productivity was caused by a superior part of O5-grades.

Shift B has more "luck". They produce thicker and wider material grades with high tonnages for automotive interior parts application purposes. The productivity of shift B considerably exceeds the productivity of shift A at the end of the month.

The same situation today:

Nowadays Shift A is still producing O5-grades for exposed automotive parts and due to their capability in every production shift possible. The average produced monthly tonnage therefore is below the values of the other shifts.

But run-time tables show that production of O5-grades takes more time than of conventional O3-grades (reduced operating speeds, more effort for plant and product quality checks, plant cleaning activities and more roll changes).

What is meant by run-time tables and nominal run-times?

For each coil a theoretical run-time is determined depending on the strip geometry (thickness and width), steel grade and final application. For calculation of the different run-times the steel works at their production sites assess themselves the operating speeds of their production facility and prove their capability. These data are stored coil wise.

Additional to the rolling speed of a coil in discontinuously operating plants further factors and parameters like material feeding in and feeding out sequences, sampling times and times for acceleration and deceleration are defined.

The sum of all these parameters gives the nominal run-time for each specific coil.

Therefore it is possible to normalize the productivity at the respective production unit (nominal run-time table) and the actual productivity in every case.

Calculation example for a hot dip galvanizing line:

A hot dip galvanizing line is operating in three shifts:

- The gross production time GPT: GPT $= 3 \cdot 8\,h = 24\,h$
- Non-productive times NPT: NPT $= 1.5\,h$
- Due to malfunctions plant production has to be stopped for two hours: MFT $= 2\,h$
- 75 coils were produced.

Therefore net operating time NOT follows according to:

NOT = GPT - (NPT + MFT)
    = 24 h - (1.5 h + 2 h)
    = 20.5 h

The plant utilization ratio UR is calculated according to:

$$UR = \frac{NOT}{GPT} \cdot 100\,\% = \frac{20.5\,h}{24\,h} \cdot 100\,\% = 85.4\,\%.$$

According to the nominal run-time table the production time for 20 coils at the hot dip galvanizing line should be 20h. In reality the production time took 20.5 h (NOT).

Summarizing the results give:
- Actual production time: 20.5h
- Deviation: 0.5h
- Utilization ratio: 85.4%

Using these numbers, the Utilization Time Factor UTF can be calculated, describing to which part the nominal productivity of the plant was fulfilled in practice by

UTF = (Nominal production time / Actual production time) x 100

UTF: $\text{UTF} = \dfrac{20\ h}{20.5\ h} \cdot 100\ \% = 97.6\ \%$.

A value characterizing to which content the available gross production time GPT was really used for production is given by the **Gross Time Factor GTF**, describing to which content the available time was used for production.

Gross time factor GTF = Utilization ratio UR x Utilization time factor UTF/100 %
GTF = 85.4 % · 97.6 % = 83.4 %

Exercise 3-2: Plant length of HSM's

In a HSM slabs with a length of 12 m, a thickness of 250 mm and a width of 1,000 mm should be rolled to a final strip with the thickness of 2.00 mm and a width of 1,000 mm. Calculate

a) the missing data in the pass schedule pattern given in **Table 3.6**,

b) the minimum plant length possible using the data from Table 3.6 for a fully continuous operating HSM with six RS's according to **Table 3.7**,

c) the minimum plant length using the data from Table 3.7 for a 3/4-continuous HSM with two reversing RS's performing three passes each, **Table 3.8**, and

d) the minimum plant length using the data from Table 3.7 for a semi-continuously operating mill with one reversing RS performing five passes and an identical transfer bar thickness as calculated in part a) of this exercise, **Table 3.9**.

Add the missing data in Tables 3.6 to 3.9. Take into consideration that no tandem rolling is allowed in the RM and the descaler and crop shear operate decoupled from the RS's.

Pass	RM/FM	Entry thickness [mm]	Exit thickness [mm]	Relative reduction [%]	Strip length [m]
1	RM	250.00		17	
2	RM		161.85		
3	RM	161.85		30	
4	RM		73.64		
5	RM	73.64		29	
6	RM			16	
7	FM	43.92	24.16		
8	FM		11.59		
9	FM			45	
10	FM	6.38		35	
11	FM		2.98		
12	FM		2.30		
13	FM	2.30		13	

Table 3.6: Pass schedule data for a final strip with the dimensions 2.00mm x 1,000mm rolled from a slab with 250mm x 1,000mm x 12m

Fully continuous HSM operating mill with 6 RS's	
Distance end furnace #1 - Furnace descaler [m]	10
Distance furnace descaler – RS R1 [m]	35.00
Distance R1-R2 [m]	
Distance R2-R3 [m]	
Distance R3-R4 [m]	
Distance R4-R5 [m]	
Distance R5-R6 [m]	
Distance R6-CS FM [m]	
Distance CS-FS F1 [m]	14.00
Length FM (F1-F7) [m]	33.00
Distance mill stand F7-DC # 1 [m]	120.00
Total length HSM [m]	

Table 3.7: Plant length of a fully continuous operating HSM with six RS's

3/4-continuously operating HSM with two reversing RS's	
Distance end furnace #1 - Furnace descaler [m]	10
Distance furnace descaler – RS R1 [m]	35.00
Distance R1-R2 [m]	
Distance R2-CS FM [m]	
Distance CS - FS F1 [m]	14.00
Length FM (F1-F7) [m]	33.00
Distance F7-DC # 1 [m]	120.00
Total length HSM[m]	

Table 3.8: Plant length of a 3/4-continuously operating HSM with two reversing RS's

Semi-continuous HSM with one reversing RS	
Distance end furnace #1 - Furnace descaler [m]	10
Distance furnace descaler - RS R1 [m]	
Distance R1 - CS FM [m]	
Distance CS - FS F1 [m]	14.00
Length FM (F1 - F7) [m]	33.00
Distance F7 - DC # 1 [m]	120.00
Total length HSM [m]	

Table 3.9: Plant length of a semi-continuously operating HSM with one reversing RS

Solution

Exercise 3-2 a

Pass	RM/FM	Entry thickness [mm]	Exit thickness [mm]	Relative reduction [%]	Strip length [m]
1	RM	250.00	207.50	17	14.46
2	RM	207.50	161.85	22	18.54
3	RM	161.85	113.30	30	26.48
4	RM	113.30	73.64	35	40.74
5	RM	73.64	52.29	29	57.38
6	RM	52.29	43.92	16	68.31
7	FM	43.92	24.16	45	124.19
8	FM	24.16	11.59	52	258.74
9	FM	11.59	6.38	45	470.43
10	FM	6.38	4.15	35	723.73
11	FM	4.15	2.98	28	1005.19
12	FM	2.98	2.30	23	1305.44
13	FM	2.30	2.00	13	1500.50

Table 3.6a: Pass schedule data for a final strip with the dimensions 2.00mm x 1,000mm rolled from a slab with 250mm x 1,000mm x 12m

The following equations can be used for calculation:

Relative thickness reduction: $\varepsilon_h = \dfrac{h_0 - h_1}{h_0}$. (3.2.1)

Strip elongation for pass no. i: $\lambda_i = \dfrac{h_{0,i}}{h_{1,i}}$. (3.2.2)

Strip length for pass i: $l_i = l_{i-1} \cdot \lambda_i$. (3.2.3)

Exercise 3-2 b

Fully continuous HSM with six RS's	
Distance end furnace #1 - Furnace descaler [m]	10
Distance furnace descaler - RS R1 [m]	35.00
Distance R1 - R2 [m]	14.46
Distance R2 - R3 [m]	18.54
Distance R3 - R4 [m]	26.48
Distance R4 - R5 [m]	40.74
Distance R5 - R6 [m]	57.38
Distance R6 - CS FM [m]	68.31
Distance CS - FS F1 [m]	14.00
Length FM (F1 - F7) [m]	33.00
Distance F7 - DC # 1 [m]	120.00
Total length HSM [m]	437.89

Table 3.7b: Plant length of a fully continuous operating HSM with six RS's

The distances in the roughing area and the length of the delay roller table have to be configured in that way that all aggregates (furnace descaler, RS's and CS) can operate decoupled from each other. Therewith, the minimum plant length of the fully continuous operating HSM takes -

according to the strip lengths calculated in Exercise 3-2 a - 438 m.

Exercise 3-2 c

3/4-continuously operating hot strip mill with two reversing RS's	
Distance end furnace #1 - Furnace descaler [m]	10
Distance furnace descaler - RS R1 [m]	35.00
Distance R1 - R2 [m]	57.38
Distance R2 - CS FM [m]	68.31
Distance CS - FS F1 [m]	14.00
Length FM (F1 - F7) [m]	33.00
Distance F7 - DC # 1 [m]	120.00
Total length hot strip mill [m]	337.69

Table 3.8c: Plant length of a 3/4-continuously operating HSM with two reversing RS's

The minimum plant length of the 3/4-continuous mill is to be calculated similar to Exercise 3-2 b. The corresponding length is 338 m.

Exercise 3-2 d

Semi-continuous HSM with one reversing RS	
Distance end furnace #1 - Furnace descaler [m]	10
Distance furnace descaler - RS R1 [m]	57.38
Distance R1 - CS FM [m]	68.31
Distance CS - FS F1 [m]	14.00
Length FM (F1 - F7) [m]	33.00
Distance F7 - DC # 1 [m]	120.00
Total length HSM [m]	302.69

Table 3.9d: Plant length of a semi-continuously operating HSM with one reversing RS

The minimum plant length of the semi-continuous mill is 303 m.

For final definition of the plant layout - besides the product mix to be rolled - investment and maintenance costs as well as targeted annual production and quality aspects have to be taken into account.

Exercise 3-3: Talking about dimensions

The maximum capable rolling force of HSM FS's is assumed to be 50 MN. How many elephants with a weight of five tons each are necessary to compensate the above given rolling force? When calculating the gravity force of the elephants use the approach $1 \text{ kg} \approx 10 \text{ N}$.

Solution:

The gravity force of an elephant is:

$$F_{Ele} = 10 \, \frac{N}{kg} \cdot 5,000 \text{ kg}$$
$$F_{Ele} = 5 \cdot 10^4 \text{ N}$$
(3.3.1)

The maximum rolling force of the stand is:

$$F_W = 50 \text{ MN}$$
$$F_W = 50 \cdot 10^6 \text{ N}$$
$$F_W = 5 \cdot 10^7 \text{ N}$$
(3.3.2)

The number of elephants for compensation of this force is:

$$n = \frac{F_W}{F_{Ele}} = \frac{5 \cdot 10^7 \text{ N}}{5 \cdot 10^4 \text{ N}} = 1 \cdot 10^3 = 1,000 \;. \tag{3.3.3}$$

To compensate the rolling force of one stand 1,000 elephants are necessary.

4. Use of a coilbox in HSM's

Exercise 4-1: Production of a hot strip mill with and without CB operation

A semi-continuously operating hot strip mill with one RS, a CB, a FM with seven stands, and three DC's is rolling 3.5 million tons of hot strip per year. A part of $A = 20\ \%$ of the production is rolled in CB operation mode without speed-up in the FM because of quality reasons or limited power in the finishing train.

Determine the possible increase of annual production after increasing the main drive power in the FM and using pass-through operation mode without CB. Take into consideration the following boundary conditions

- Average slab width: $B = 1{,}250$ mm,
- Average slab thickness: $H = 250$ mm,
- Average slab length: $L = 9$ m,
- Average final strip width: $b = 1{,}250$ mm
- Average final strip thickness: $h = 3.75$ mm,
- Average density of the rolled steel grade: $\rho = 7.55\ \text{t m}^{-3}$,
- Speed-up value (acceleration): $a = 0.2\ \text{m s}^{-2}$ (rolling without CB),
- Distance F7 - DC: $l_0 = 100$ m.

Solution:

According to the mass flow law together with the given production of $P = 3{,}500{,}000$ t/a the average final rolling speed using the CB is given by

$$V_0 = \frac{P}{\rho \cdot b \cdot h}$$

$$V_0 = \frac{3{,}500{,}000 \text{ t}}{365 \cdot 24 \cdot 3{,}600 \text{ s}} \cdot \frac{1 \text{ m}^3}{7.55 \text{ t}} \cdot \frac{1}{1.250 \text{ m} \cdot 0.00375 \text{ m}} \quad (4.1.1)$$

$$V_0 = 3.14 \, \frac{\text{m}}{\text{s}}$$

Following the law of mass flow $B \cdot H \cdot L = b \cdot h \cdot l$ the average final strip length l can be calculated

$$l = \frac{B \cdot H}{b \cdot h} \cdot L$$

$$l = \frac{1{,}250 \text{ mm} \cdot 250 \text{ mm}}{1{,}250 \text{ mm} \cdot 3.75 \text{ mm}} \cdot 9 \text{ m} = 600 \text{ m} \quad (4.1.2)$$

The average starting time for down coiling of the final strip leaving the finishing train is

$$t_0 = \frac{l_0}{V_0}$$

$$t_0 = \frac{100}{3.14} \text{ s} = 32 \text{ s} \quad (4.1.3)$$

The average total strip running time from the last FS to the DC is

$$t = \frac{l}{V_0}$$

$$t = \frac{600 \text{ m}}{3.14 \text{ m s}^{-1}} = 191 \text{ s} \quad (4.1.4)$$

Using the main drive acceleration operation technique (speed-up) the resulting strip travelling time t_R follows according to Chapter 5 of the book "Basics in flat rolling" /27/.

$$l - l_0 = v_0 \cdot t_R + \frac{1}{2} \cdot a \cdot t_R^2$$

$$\Leftrightarrow t_R = -\frac{v_0}{a} + \sqrt{\left(\frac{v_0}{a}\right)^2 + \frac{2 \cdot (l - l_0)}{a}} \quad . \tag{4.1.5}$$

$$\Rightarrow t_R = -15{,}7 \text{ s} + \sqrt{246.49 + 5000} \text{ s}$$

$$t_R = 57 \text{ s}$$

The total strip rolling time t_{su} of the coils produced with speed-up therefore is

$$t_{su} = t_0 + t_R$$
$$t_{su} = 32 \text{ s} + 57 \text{ s} \,. \tag{4.1.6}$$
$$t_{su} = 89 \text{ s}$$

The annual increase in hot strip mill productivity can be derived from the ratio of the total strip running times with and without CB/speed-up operation

$$\left(\frac{t}{t_{su}} - 1\right) \cdot P \cdot A = \left(\frac{191 \text{s}}{89 \text{s}} - 1\right) \cdot 3{,}500{,}000 \, \frac{t}{a} \cdot 0{,}2 \approx 810.000 \, \frac{t}{a} \tag{4.1.7}$$

or 23%, **Fig. 4.1** and **4.2**.

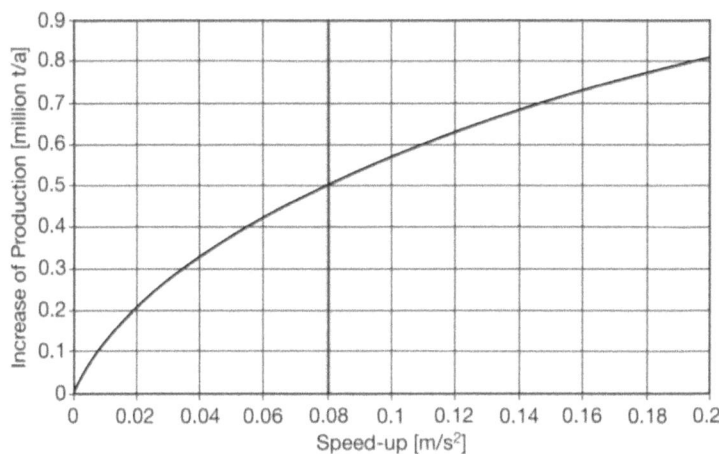

Fig. 4.1: Production increase of a semi-continuous HSM as a function of speed-up rates (Basic production: 3.5 million t/a), results absolute. The red line at the speed-up value of 0.08 m/s^2 gives the transition from normal to power temperature speed-up /26/

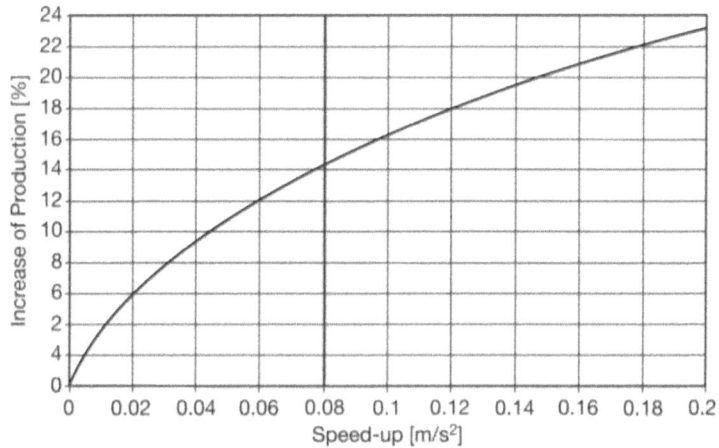

Fig. 4.2: Production increase of a semi-continuous HSM as a function of speed-up rates (Basic production: 3.5 million t/a), results relative. The red line at the speed-up value of 0.08 m/s^2 gives the transition from normal to power temperature speed-up /26/

Exercise 4-2: Rolling in CB mode – Reduction of strip surface radiation area

A transfer bar with a length l = 70 m, width b = 1,600 mm and thickness h = 35 mm is rolled in CB mode. The inner diameter of the coil is $d_i = 600$ mm. Determine the radiation area of the uncoiled transfer bar and compare it to the radiation area of the coiled transfer bar. Calculate their ratio.

Solution:

The radiation area A_{TB} of the uncoiled transfer bar is:

$A_{TB} = 2 \cdot (l \cdot b + l \cdot h + b \cdot h)$
$A_{TB} = 2 \cdot (70 \text{ m} \cdot 1.6 \text{ m} + 70 \text{ m} \cdot 0.035 \text{ m} + 1.6 \text{ m} \cdot 0.035 \text{ m})$. (4.2.1)
$A_{TB} = 229 \text{ m}^2$

The outer diameter d_o of the coiled transfer bar can be calculated using the transfer bar length and thickness and inner coil diameter, **Fig. 4.3**, according to equation

$l \cdot h = \dfrac{d_o^2 - d_i^2}{4} \cdot \pi$

$\Leftrightarrow d_o = \sqrt{\dfrac{4 \cdot l \cdot h + d_i^2}{\pi}}$. (4.2.2)

Inserting gives for the outer diameter:

$d_o = 1.798 \text{ m}$. (4.2.3)

The outside radiation area of the coiled transfer bar therewith is:

$$A_{CB} = \frac{\pi}{2} \cdot (d_o^2 - d_i^2) + \pi \cdot d_o \cdot b \quad (4.2.4)$$
$$A_{CB} = 14 \text{ m}^2$$

Therefore the ratio of both radiation areas is:

$$\frac{A_{CB}}{A_{TB}} = \frac{14 \text{ m}^2}{229 \text{ m}^2} = 6\% . \quad (4.2.5)$$

As a result, temperature loss by radiation can be reduced approximately 17 times using a CB!

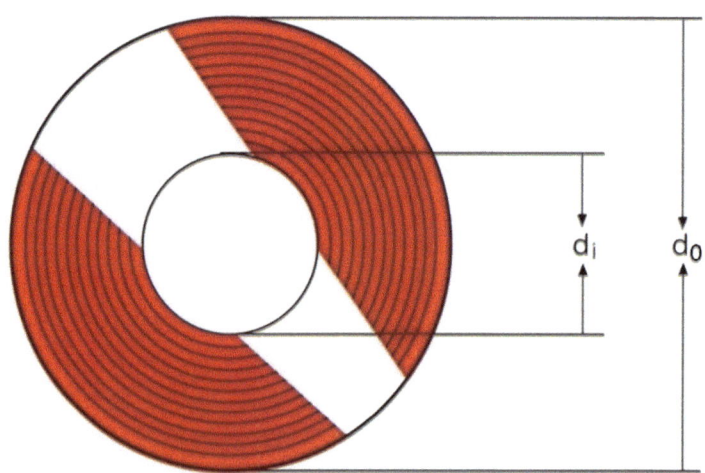

Fig. 4.3: Scheme for coiled transfer bar; determining the outer coil diameter d_o

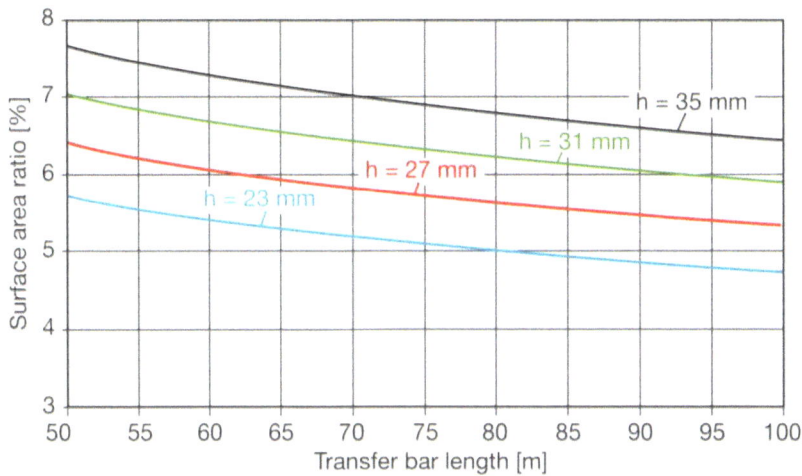

Fig. 4.4: Ratio of the radiation areas with/without CB of the transfer bar (width b=1,000mm), Parameter: transfer bar thickness

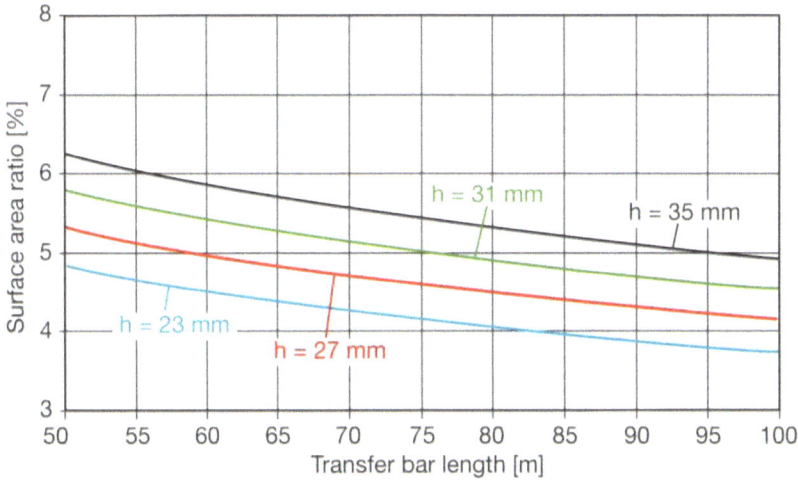

Fig. 4.5: Ratio of the radiation areas with/without CB of the transfer bar (width b=2,000mm), Parameter: transfer bar thickness

Fig. 4.4 and **4.5** show the ratio A_{CB}/A_{TB} of the radiation areas for the transfer bar widths 1,000 mm and 2,000 mm as function of the transfer length for several transfer bar thicknesses. For constant transfer bar width and thickness the ratio-values depend only slightly from the transfer bar length. It takes values between 3.7 % and 7.6 % according to chosen transfer bar width and thickness.

5. Process automation in HSM's

Exercise 5-1: Roll gap geometry with biting condition and thickness reduction

The transfer bar thickness in a HSM is $h_0 = 44$ mm and the friction coefficient in the roll gap is $\mu = 0.22$. The work roll radius of F1 is $r = 380$ mm. Calculate the limit value for the absolute and relative thickness reduction for the rolling pass so that the biting conditions are fulfilled.

Solution:

Due to the biting condition derived in Chapter 3 of the book "Basics in flat rolling" /27/ $\tan \alpha_0 < \mu$ the biting angle follows with $\alpha_0 = 12.4°$.

The absolute thickness draft can be calculated following

$$\cos \alpha_0 = 1 - \frac{h_0 - h_1}{2 \cdot r} \tag{5.1.1}$$

and

$$\begin{aligned} h_0 - h_1 &= 2 \cdot r \cdot (1 - \cos \alpha_0) \\ h_0 - h_1 &= 760 \text{ mm} \cdot (1 - \cos 12.4°) = 17.7 \text{ mm} \end{aligned} \tag{5.1.2}$$

The relative thickness draft follows according to

$$\varepsilon_h = \frac{h_0 - h_1}{h_0} = \frac{17.7 \text{ mm}}{44 \text{ mm}} = 40\ \%. \tag{5.1.3}$$

Exercise 5-2: Roll force and yield strength

Pass schedule calculation in a HSM with six FS's is performed using Sims roll gap model /27/. Calculate with the given pass schedule data the average yield strength of the strip in the FS. The transfer bar thickness is 44 mm, the final strip thickness is 6.24 mm and the strip width is 1,600 mm. Complete the missing pass schedule data in **Table 5.1**. The slip forward factor of the strip can be neglected when calculating roll peripheral speed.

FS	F1	F2	F3	F4	F5	F6
Roll radius r [mm]	370	370	370	370	370	370
Exit thickness h_1 [mm]	30.83	18.72	13.44	10.28	7.49	6.24
Absolute thickness draft [mm]						
Relative thickness draft [-]						
Roll gap length [mm]						
Roll peripheral speed [m/s]	0.50					
Logarithmic deformation [-]						
Deformation rate [1/s]						
Geometry factor Qp according to Sims roll gap model [-]	1.19	1.49	1.56	1.47	1.80	1.70
Roll force F [kN]	19,500	18,000	13,600	12,600	16,000	9,900
Yield strength k_{fm} [N/mm²]						

Table 5.1: Pass schedule data FM for steel grade C15 with a final strip thickness of 6.24mm

Solution:

Calculations can be performed using equations as follows /27/:

Absolute thickness draft: $\Delta h = h_0 - h_1$, (5.2.1)

Relative thickness draft: $\varepsilon_h = \dfrac{h_0 - h_1}{h_0} = 1 - \dfrac{h_1}{h_0}$. (5.2.2)

Roll bite length: $l_d = \sqrt{r \cdot (h_0 - h_1)} = \sqrt{r \cdot \Delta h}$. (5.2.3)

Roll peripheral speed: $v_U \approx v_1 = \dfrac{h_0}{h_1} \cdot v_0$, (5.2.4)

Logarithmic deformation: $\varphi = \ln\left(\dfrac{h_0}{h_1}\right) = \ln(1+\varepsilon)$ (5.2.5)

Deformation rate: $\dot{\varphi} = \dfrac{v_U}{l_d} \cdot \varphi$. (5.2.6)

Roll force according to Sims: $F = k_{fm} \cdot l_d \cdot b \cdot Q_P\left(\dfrac{r}{h_1}, \varepsilon\right)$ (5.2.7)

k_{fm}: Average yield strength; $Q_P = f\left(\dfrac{r}{h_1}, \varepsilon_h\right)$ geometry factor.

The arrangement of the roll force formula provides the average yield strength:

$$k_{fm} = \dfrac{F}{l_d \cdot b \cdot Q_P\left(\dfrac{r}{h_1}, \varepsilon_h\right)}.$$ (5.2.8)

The solutions are:

FS	F1	F2	F3	F4	F5	F6
Absolute thickness draft [mm]	13.17	12.11	5.28	3.16	2.79	1.25
Relative thickness draft [-]	0.30	0.39	0.28	0.24	0.27	0.17
Roll gap length l_d [mm]	69.81	66.94	44.20	34.19	32.13	21.51
Roll peripheral speed v_u [m/s]	0,50	0.82	1.15	1.50	1.70	2.04
Logarithmic deformation [-]	0,26	0,33	0,25	0,22	0,24	0,16
Deformation rate [1/s]	1.86	4.04	6.50	9.65	12.70	15.17
Yield strength k_{fm} [N/mm²]	147	113	123	157	173	169

Table 5.2: Pass schedule missing data for steel grade C15 with a final strip thickness of 6.24mm

Exercise 5-3: Lever arm method and roll torque in HSM's

In a HSM the following pass schedule data are given: Entry thickness $h_0 = 36$ mm, relative thickness draft $\varepsilon_h = 44.56\,\%$, work roll radius $r = 325$ mm, rolling force $F = 14,297$ kN and the lever arm /27/ $m = 0.452$ mm. Calculate the roll torque using the lever arm method according to Trinks.

Solution:

The formula for the calculation of the roll torque for hot rolling - with working rolls of the same diameter (symmetric rolling) -according to Trinks is:

$$M = 2 \cdot F \cdot l_d \cdot m \qquad (5.3.1)$$

with

$$l_d = \sqrt{r \cdot (h_0 - h_1)} \qquad (5.3.2)$$

as roll bite length (contact length) and h_1 as exit thickness. The lever arm m describes the distance of the vertical roll force from the roll rotating point. The exit thickness can be calculated according to

$$h_1 = h_0 \cdot (1-\varepsilon_h) = 36 \text{ mm} \cdot (1-0.4456) = 19.96 \text{ mm}. \quad (5.3.3)$$

Using the values given before provides the roll torque:

$$M = 2 \cdot 14,297 \text{ kN} \cdot \sqrt{325 \text{ mm} \cdot (36 \text{ mm} - 19.96 \text{ mm})} \cdot 0.452 \quad (5.3.4)$$
$$M = 933 \text{ kNm}$$

Remarks due to heat balance during hot rolling

Questions of temperature calculations during hot rolling are treated in Exercises 5-4 and 5-5. The necessary formula work for solving the exercises are given in Chapter 16 in"Basics in flat rolling" /27/.

Exercise 5-4: Temperature calculation for slab rolling in the RM

A slab with the dimensions $h_0 = 230$ mm, $b_0 = b_1 = 1,500$ mm and $l_0 = 11,000$ mm is discharged from the reheating furnace of the HSM with a temperature of $T_M = 1,250$ °C. The transfer time on the roller table to the first RS takes t_t = 88 s. The exit thickness of the slab after the first pass is $h_1 = 195$ mm.

For the rolling pass the further data are known:

- Deformation stress $k_w = 60$ N mm^{-2},
- Work roll diameter $d = 1,200$ mm,
- Exit speed of the slab $v_1 = 1.50$ m s^{-1},
- Core temperature of the work roll $T_C = 55$ °C,
- Ambient temperature $T_A = 25$ °C.

Calculate the following parameters for a heavy scaled slab surface with an emissivity factor 0.85 neglecting slab spreading and slip forward speed factor:

a.) Temperature loss due to heat radiation for slab head and tail end for the respective pass.

b.) Temperature loss due to heat conduction to the rolls for the slab tail end.

c.) Temperature increase due to the deformation energy (dissipation).

d.) Determine the temperature for the slab tail end leaving the mill stand.

Solution:

Exercise 5-4-a

The temperature loss due to heat radiation for the slab head end is given by the formula (16.2) in /27/:

$$\Delta T_{rad} = \frac{\varepsilon \cdot C_S \cdot t_t \cdot U_S}{A_S \cdot \rho \cdot c_P} \cdot \left[\left(\frac{T_M}{100} \right)^4 - \left(\frac{T_A}{100} \right)^4 \right]. \qquad (16.2)$$

Density and specific heat capacity both depend on temperature. These data can be taken out of tables or can be calculated for basic carbon grades rolled within the temperature window between 800°C and 1,300°C according to the equations

- Specific heat capacity:
  $c_P = c_P(T) = 454.54 + 0.327 \cdot T$ [J kg^{-1} K^{-1}],
- Material density:
  $\rho = \rho(T) = 8.0332 \cdot 10^{-6} - 4.833 \cdot 10^{-10} \cdot T$ [kg mm^{-3}].

Temperatures have to be inserted in [°C].

When discharging the slab from the furnace with the temperature of 1,250°C the following data can be calculated:

$$c_P = 454.54 \text{ J kg}^{-1} \text{ K}^{-1} + 0.327 \cdot 1,250 \text{ J kg}^{-1} \text{ K}^{-1}$$
$$c_P = 863.29 \text{ J kg}^{-1} \text{ K}^{-1} \qquad (5.4.1)$$

$$\rho = 8.0332 \cdot 10^{-6} \text{ kg mm}^{-3} - 4.833 \cdot 10^{-10} \cdot 1,250 \text{ kg mm}^{-3}$$
$$\rho = 7.429 \cdot 10^{-6} \text{ kg mm}^{-3} \qquad (5.4.2)$$

The slab's circumference is

$$U_0 = 2 \cdot (h_0 + b_0)$$
$$= 2 \cdot (230 \text{ mm} + 1,500 \text{ mm}), \qquad (5.4.3)$$
$$= 3,460 \text{ mm}$$

and the slab's cross section is calculated according to:

$$A_0 = b_0 \cdot h_0 = 1,500 \text{ mm} \cdot 230 \text{ mm}$$
$$A_0 = 345,000 \text{ mm}^2 \qquad (5.4.4)$$

The absolute slab and ambient temperatures are: $T_M = 1,523 \text{ K}$, $T_A = 298 \text{ K}$.

The emissivity factor for a heavy scaled surface is given with $\varepsilon = 0.85$.

Inserting the data in equation (16.1) gives the temperature loss due to the heat radiation at the slab head end.

$$\Delta T_{rad} = \frac{0.85 \cdot 5.786 \cdot 88 \cdot 3,460}{345,000 \cdot 7.429 \cdot 863.29} \cdot \left[15.23^4 - 2.98^4\right] \frac{\text{J s mm}^4 \text{ kg K}^5}{\text{mm}^4 \text{ s K}^4 \text{ kg J}} . (5.4.5)$$
$$\Delta T_{rad} = 34 \text{ K}$$

The temperature loss of the slab head due to heat radiation before entering the rougher amounts 34 K.

The same consideration has to be done for the slab end. It has to be taken into account that the slab rests before the rolling stand while the other slab part is rolled. The relevant difference in time is the rolling time t_r.

The rolling time t_r is:

$$t_r = \frac{l_1}{v_1} . \qquad (5.4.6)$$

The exit slab length can be determined using the law of volume constancy:

$$l_0 \cdot b_0 \cdot h_0 = l_1 \cdot b_1 \cdot h_1$$

$$l_1 = \frac{l_0 \cdot b_0 \cdot h_0}{b_1 \cdot h_1}$$

$$l_1 = \frac{11{,}000 \text{ mm} \cdot 1{,}500 \text{ mm} \cdot 230 \text{ mm}}{1{,}500 \text{ mm} \cdot 195 \text{ mm}}$$

$$l_1 = 12{,}974 \text{ mm}$$

(5.4.7)

Therewith the rolling time can be calculated to

$$t_r = \frac{l_1}{v_1} = \frac{12{,}974 \text{ mm}}{1{,}250 \text{ mm s}^{-1}} = 10.38 \text{ s} \approx 10 \text{ s}.$$

(5.4.8)

The total time from the slab leaving the furnace until the slab end entering the roll gap of the RM amounts to:

$$t_{tot} = t_r + t_t$$

$$t_{tot} = 10.38 \text{ s} + 88 \text{ s} = 98.38 \text{ s} \approx 98 \text{ s}$$

(5.4.9)

For the radiation loss still (16.2) is valid:

$$\Delta T_{rad} = \frac{\varepsilon \cdot C_S \cdot t_t \cdot U_S}{A_S \cdot \rho \cdot c_P} \cdot \left[\left(\frac{T_M}{100}\right)^4 - \left(\frac{T_A}{100}\right)^4 \right]. \qquad (16.2)$$

The time to be inserted in the equation is $t_{tot} = 98$ s. All other data remain unchanged and can be taken analogously of the slab head end consideration. The temperature loss due to radiation for the slab end therefore is:

$$\Delta T_{rad} = \frac{0.85 \cdot 5.786 \cdot 98 \cdot 3,460}{345,000 \cdot 7.429 \cdot 863.29} \cdot \left[15.23^4 - 2.98^4\right] \frac{J\, s\, mm^4\, kg\, K^5}{mm^4\, s\, K^4\, kg\, J}. \quad (5.4.10)$$

$\Delta T_{rad} = 38$ K

The slab end loses 38K temperature until the pass begins. In comparison to the slab head the difference amounts $\Delta T = 38\,K - 34\,K = 4\,K$.

<u>Exercise 5-4-b</u>

For calculation of the temperature loss due to heat conduction the equation (16.6) from Chapter 16 in "Basics in flat rolling" /27/ is valid:

$$\Delta T_L = \frac{2 \cdot l_d \cdot b_m \cdot k_L \cdot (T_M - T_C)}{c_P \cdot \rho \cdot \dot{V}}. \qquad (16.6)$$

The roll gap length can be calculated according to:

$$l_d = \sqrt{r \cdot (h_0 - h_1)}$$
$$l_d = \sqrt{600 \text{ mm} \cdot (230 \text{ mm} - 195 \text{ mm})} = 145 \text{ mm} \quad (5.4.11)$$

The average slab width when neglecting spreading is:

$$b_m = b_0 = b_1 = 1{,}500 \text{ mm}. \quad (5.4.12)$$

The heat transfer coefficient can be calculated according to Pawelski formula /27/. It depends on the work roll contact time of the slab and the scale thickness. The contact time is:

$$t_c = \frac{l_d}{v_1}$$
$$t_c = \frac{145 \text{ mm}}{1{,}500 \text{ mm} \cdot \text{s}^{-1}} \quad (5.4.13)$$
$$= 9{,}66 \cdot 10^{-2} \text{ s}$$
$$\approx 10 \text{ ms}$$

Therefore the heat transfer coefficient for heavy scale formation on the slab can be calculated according to equation (16.7) given in /27/:

$$k_{Lo} = \frac{8.267 \cdot 10^{-6}}{\sqrt{t_c}}$$
$$k_{Lo} = 2.660 \cdot 10^{-5} \; \frac{\text{kJ}}{\text{mm}^2 \text{ s K}} \quad (16.7)$$

The volume flow for the slab end is:

$$\dot{V} = A_1 \cdot v_1$$

$$\dot{V} = 1{,}500 \text{ mm} \cdot 195 \text{ mm} \cdot 1.5 \cdot 10^3 \text{ mm s}^{-1}. \quad (5.4.14)$$

$$\dot{V} = 438.75 \cdot 10^6 \text{ mm s}^{-1}$$

Specific heat capacity and material density have to be calculated again because of the altered starting temperature (the colder slab end enters now the rolling stands). The relevant temperature then is

$$T_1 = T_0 - T_{rad}$$
$$T_1 = 1{,}250 \text{ °C} - 38 \text{ °C} = 1{,}212 \text{ °C} \quad (5.4.15)$$

and therewith:

$$c_P = 454{,}54 \text{ J kg}^{-1} \text{ K}^{-1} + 0.327 \cdot 1{,}212 \text{ J kg}^{-1} \text{ K}^{-1}$$
$$c_P = 851 \text{ J kg}^{-1} \text{ K}^{-1} \quad (5.4.16)$$

$$\rho = 8{,}03 \cdot 10^{-6} \text{ kg mm}^{-3} - 4{,}83 \cdot 10^{-10} \cdot 1{,}21 \text{ kg mm}^{-3}$$
$$\rho = 7.45 \cdot 10^{-6} \text{ kg mm}^{-3} \quad (5.4.17)$$

For the temperature loss due to heat conduction still equation (16.6) is valid:

$$\Delta T_L = \frac{2 \cdot l_d \cdot b_m \cdot k_L \cdot (T_M - T_C)}{c_P \cdot \rho \cdot \dot{V}}. \quad (16.6)$$

Inserting all given and calculated data the temperature loss due to heat conduction at the slab end is given by:

$$\Delta T_L = 4.8\,K \approx 5\,K. \tag{5.4.18}$$

In the roll gap the slab end loses 5K due to the contact with the rolls.

Exercise 5-4-c

Besides the temperature losses due to heat radiation and thermal conductivity an increase of slab temperature is created due to the dissipation energy by deformation.

The associated temperature increase can be calculated as

$$\Delta T_{DEF} = \frac{k_w \cdot \varphi_V}{c_p \cdot \rho}. \tag{5.4.19}$$

with φ_V the effective strain /27/. According to the von Mises theory the effective strain can be calculated with the corresponding main strains

$$\varphi_V = \sqrt{\frac{2}{3} \cdot \left(\varphi_h^2 + \varphi_b^2 + \varphi_l^2\right)}. \tag{5.4.20}$$

The main strain in thickness direction can be determined with the entry and exit strip thickness according to:

$$\varphi_h = \ln\left(\frac{h_1}{h_0}\right)$$

$$\varphi_h = \ln\left(\frac{230\,mm}{195\,mm}\right) = -0.165 \tag{5.4.21}$$

Using the volume constancy condition:

$$\varphi_h + \varphi_b + \varphi_l = 0. \tag{5.4.22}$$

Assuming no spread, the main strain in width direction is $\varphi_b = 0$, following:

$$\varphi_h + \varphi_l = 0$$
$$\varphi_l = -\varphi_h = +0.165 \tag{5.4.23}$$

Therewith the effective strain can be calculated according to

$$\varphi_V = \sqrt{\frac{2}{3} \cdot \left((-0.165)^2 + 0^2 + (0.165)^2\right)}. \tag{5.4.24}$$
$$\varphi_V = 0.191$$

All other data are known from the given exercise. Therefore, together with the calculated specific heat capacity and material density from exercise b), the temperature increase due to dissipation is:

$$\Delta T_{DEF} = \frac{k_w \cdot \varphi_V}{c_p \cdot \rho}. \tag{5.4.25}$$
$$\Delta T_{DEF} = 2 \text{ K}$$

The slab temperature increase in the roll gap is 2 K.

Exercise 5-4-d

Finally all calculated temperature data are summarized. The complete temperature balance can be written according to:

$$T_E = T_M - \Delta T_{rad} - \Delta T_L + \Delta T_{DEF}$$
$$T_E = 1{,}250\ °C - 38\ °C - 5\ °C + 2\ °C = 1{,}209\ °C \quad (5.4.26)$$

The temperature of the slab's tail end leaving the RM stand therefore amounts to 1,209°C; the temperature loss is 41 K.

Exercise 5-5: Run-time of transfer bars

A transfer bar with the thickness h = 40 mm and the width b = 1500 mm exits the RM with the temperature of $T_M = 1{,}090\ °C$ at its head end. Its entry temperature at the CS is 1,043°C and the ambient temperature is 25°C.

Calculate the run-time of the transfer bar's head end to the CS taking into account the average material density of $\rho = 7.506 \cdot 10^{-6}\ kg\ mm^{-3}$ and average specific heat capacity of $c_p = 810\ J\ kg^{-1}\ K^{-1}$.

Determine the average velocity of the transfer bar head end on the delay roller table to the CS. The distance of the last RS to the CS is 90 m.

Use $\varepsilon = 0.80$ as emissivity factor for the scaled transfer bar surface.

Solution:

The temperature loss of the transfer bar's head end /27/ is given in analogy to Exercise 5-4 by:

$$\Delta T_{rad} = \frac{\varepsilon \cdot C_s \cdot t_t \cdot U_s}{A_s \cdot \rho \cdot c_p} \cdot \left[\left(\frac{T_M}{100}\right)^4 - \left(\frac{T_A}{100}\right)^4\right]. \tag{16.2}$$

Material density and specific heat capacity are as given in the exercise.

Calculation of the transfer bar's circumference:

$$U_s = 2 \cdot (h+b)$$
$$U_s = 2 \cdot (40 \text{ mm} + 1{,}500 \text{ mm}) = 3{,}080 \text{ mm} \tag{5.5.1}$$

The cross section of the transfer bar is:

$$A_s = b \cdot h$$
$$A_s = 1{,}500 \text{ mm} \cdot 40 \text{ mm} = 60{,}000 \text{ mm}^2 \tag{5.5.2}$$

The absolute temperatures for the transfer bar and the surrounding area are $T_M = 1{,}523$ K and $T_A = 298$ K, respectively.

Rearrangement of the temperature loss equation and inserting the given and calculated parameters provides $t_t = 35$ s for the run-time of the transfer bar's head end to the CS.

With the given distance of 90 m between RM and CS the average speed of the transfer bar's head end is 2.57 m/s.

By implementation of heat cover shields on the delay roller table the temperature loss of the transfer bar due to heat radiation can be reduced strongly, so increasing the entry temperature into the FM and the temperature wedge between head and tail end of the transfer bar reaching the first stand of the FM is dropped down. In consequence, speed-up is reduced.

The two photos in **Fig. 5.2** show one type of heat cover shields on the delay roller table in open and closed position. Another possibility to increase transfer bar temperature is the implementation of high-capacity mill stand main drives in the RM with speeds up to 7 m/s, further reducing run-times of slabs and transfer bars.

Fig. 5.2: Heat cover shields in open (left) and closed (right) position. Photos: SMS group GmbH

Nevertheless, speed-up is necessary to overcome temperature loss of the bar from head to tail end, laying in front of the FM. Therefore a CB or an active heating cover shield can be used to avoid this effect completely.

Exercise 5-6: Hot flow curves and hot compression tests

For three steel grades a hot flow curve /27/ is given according to

$$k_f\left(\varphi, \dot{\varphi}, T\right) = K \cdot \exp(m_1 \cdot T) \cdot \varphi^{m_2 + m_5 \cdot T} \cdot \dot{\varphi}^{m_3} \cdot \exp(m_4 \cdot \varphi). \quad (5.6.1)$$

The coefficients for the grades are listed in **Table 5.3**:

Steel grade	K	m_1	m_2	m_3	m_4	m_5
C15	5050	-0.0031256	-0.16862	0.35921	-0.70212	0.00027833
X210Cr12	15819	-0.0037549	0.0077829	0.22806	-1.052	0.00010761
NiCr80.20	218380	-0.0041074	-0.20484	0.22954	-0.58457	0.00031476

Table 5.3: Flow curve parameter

a.) Calculate the flow stress for the grades for the parameters

(I) $\varphi = 0.2$; $\dot{\varphi} = 1\,\text{s}^{-1}$; $T = 900\,°C$,

(II) $\varphi = 0.5$; $\dot{\varphi} = 50\,\text{s}^{-1}$; $T = 1050\,°C$,

(III) $\varphi = 0.7$; $\dot{\varphi} = 100\,\text{s}^{-1}$; $T = 1150\,°C$.

b.) Compression tests with cylindrical samples with the dimensions $d_0 = 25\,\text{mm}$, $h_0 = 50\,\text{mm}$ are carried out with a constant pressing tool speed of (I) $v_{Pt} = 50\,\text{mm s}^{-1}$ and (II) $v_{Pt} = 2000\,\text{mm s}^{-1}$ respectively.

Determine the deformation rate $\dot{\varphi}_0$ at the beginning and $\dot{\varphi}_1$ at the end of the compression test for the sample heights (I) h_1= 30mm and (II) h_2= 15 mm.

c.) Calculate the average deformation rate for the forming steps according to part b.

Solution:

Exercise 5-6-a

The solutions for the three steel grades can be calculated according to the flow curve

$$k_f\left(\varphi,\dot{\varphi},T\right)=K\cdot\exp(m_1\cdot T)\cdot\varphi^{m_2+m_5\cdot T}\cdot\dot{\varphi}^{m_3}\cdot\exp(m_4\cdot\varphi). \quad (5.6.1)$$

Inserting the given constants for the three steel grades provide:

Steel grade C15

$$k_f\left(\varphi,\dot{\varphi},T\right)=5050\cdot\exp(-0.00331256\cdot T)\cdot\varphi^{-0.16862+0.00027833\cdot T}$$
$$\cdot\dot{\varphi}^{0.35921}\cdot\exp(-0.70212\cdot\varphi)$$

Steel grade X210Cr12

$$k_f\left(\varphi,\dot{\varphi},T\right)=15819.0\cdot\exp(-0.0037549\cdot T)\cdot\varphi^{0.0077829+0.00010761\cdot T}$$
$$\cdot\dot{\varphi}^{0.22806}\cdot\exp(-1.052\cdot\varphi)$$

Steel grade NiCr80.20

$$k_f\left(\varphi,\dot{\varphi},T\right)=21838\cdot\exp(-0.0041074\cdot T)\cdot\dot{\varphi}^{-0.20484+0.00031476\cdot T}$$
$$\cdot\varphi^{0.22954}\cdot\exp(-0.584572\cdot\varphi).$$

The resulting yield stress values are listed in **Table 5.4**.

Steel Grade	Parameter			Yield stress [N mm^{-2}]
	φ [-]	$\dot{\varphi}$ [s^{-1}]	T [°C]	
C15	0.2	1	900	333
C15	0.5	50	1050	305
C15	0.7	100	1150	245
X210Cr12	0.2	1	900	303
X210Cr12	0.5	50	1050	248
X210Cr12	0.7	100	1150	171
NiCr80.20	0.2	1	900	333
NiCr80.20	0.5	50	1050	305
NiCr80.20	0.7	100	1150	245

Table 5.4: Yield stress data

Exercise 5-6-b

The deformation rate for constant pressing tool speed is:

$$\dot{\varphi}=\frac{d\varphi}{dt}=\frac{v_{Pt}}{h(t)}. \qquad (5.6.2)$$

For $v_{Pt}=50$ mm/s as pressing tool speed it follows

$$\dot{\varphi}_0=\frac{50\text{ mm}}{50\text{ s mm}}=1\text{ s}^{-1}$$

for the deformation rate at the beginning and for the two final heights at the end of the compression test:

$$\dot{\varphi}_1 = \frac{50 \text{ mm}}{30 \text{ s mm}} = 1{,}67 \text{ s}^{-1} \qquad \text{for } h_1 = 30 \text{ mm},$$

$$\dot{\varphi}_1 = \frac{50 \text{ mm}}{15 \text{ s mm}} = 3{,}33 \text{ s}^{-1} \qquad \text{for } h_1 = 15 \text{ mm}.$$

For $v_{Pt} = 2000 \text{ mm s}^{-1}$ as pressing tool speed it follows by analogy

$$\dot{\varphi}_1 = \frac{2000 \text{ mm}}{30 \text{ s mm}} = 66.67 \text{ s}^{-1} \qquad \text{for } h_1 = 30 \text{ mm},$$

$$\dot{\varphi}_1 = \frac{2000 \text{ mm}}{15 \text{ s mm}} = 133.33 \text{ s}^{-1} \qquad \text{for } h_1 = 15 \text{ mm}.$$

Exercise 5-6-c

The average deformation rate is the average of the two deformation rates at the start and end of the compression test. Therefore it is:

$v_{Pt} = 50 \text{ mm s}^{-1}$

$$\dot{\varphi}_m = \frac{\dot{\varphi}_0 + \dot{\varphi}_1}{2} \qquad \text{for } h_1 = 30 \text{ mm},$$

$$\dot{\varphi}_m = \frac{1.67 \text{ s}^{-1} + 1.0 \text{ s}^{-1}}{2} = 1.335 \text{ s}^{-1}$$

$$\dot{\varphi}_m = \frac{\dot{\varphi}_0 + \dot{\varphi}_1}{2}$$

$$\dot{\varphi}_m = \frac{3.33 \text{ s}^{-1} + 1.0 \text{ s}^{-1}}{2} = 2.17 \text{ s}^{-1}$$

for $h_1 = 15$ mm.

and for

$v_{Pt} = 2000$ mm s^{-1}

$$\dot{\varphi}_m = \frac{\dot{\varphi}_0 + \dot{\varphi}_1}{2}$$

$$\dot{\varphi}_m = \frac{40 \text{ s}^{-1} + 66.67 \text{ s}^{-1}}{2} = 53.33 \text{ s}^{-1}$$

for $h_1 = 30$ mm,

$$\dot{\varphi}_m = \frac{\dot{\varphi}_0 + \dot{\varphi}_1}{2}$$

$$\dot{\varphi}_m = \frac{40 \text{ s}^{-1} + 133.33 \text{ s}^{-1}}{2} = 86.67 \text{ s}^{-1}$$

for $h_1 = 15$ mm.

Numerical integration provides more exact results than simple averaging the start and end values of the deformation rate. Exact calculation of hot compression including average k_{fm}-values and deformation energy is as well only possible using numerical methods. The temperature increase due to deformation heat can be then incorporated in the calculations.

6. Thickness control in HSM's

Exercise 6-1: Hysteresis mill stand modulus

In a HSM the mill stands modulus is $C_G = 7,000$ kN mm^{-1}. The hysteresis at the AGC operating point is 0.030 mm. Derive the general equation for calculation the exit thickness error for a given hysteresis and no variation of deformation resistance and entry thickness defect. Calculate the resulting thickness error for the material modulus $C_M = 7,000$ kN mm^{-1}.

Solution:

From the general relation for calculating the error /32/

$$dh_1 = \left(\frac{dk_w}{k_w} \cdot 2 \cdot \Delta h + dh_0 + ds \cdot \frac{C_G}{C_M} \right) \cdot \frac{C_M}{C_M + C_G} \qquad (6.1.1)$$

with

k_w : Deformation resistance of the strip,
dk_w : Variation deformation resistance caused by material or temperature error,
Δh : Absolute thickness draft,
dh_0 : Entry thickness error,
ds : Errors in the mill stand (e.g. roll eccentricities, roll thermal expansion, hysteresis),
C_M : Material modulus,
C_G : Mill stand modulus

Follows with $dk_W = dh_0 = 0$:

$$dh_1 = \left(ds \cdot \frac{C_G}{C_M}\right) \cdot \frac{C_M}{C_M + C_G} = ds \cdot \frac{C_G}{C_M + C_G}. \qquad (6.1.2)$$

With the given values:

$$dh_1 = 0.030 \text{ mm} \cdot \frac{7,000}{7,000 + 7,000} \frac{\text{kN}}{\text{mm}}$$
$$dh_1 = 0.015 \text{ mm} \qquad (6.1.3)$$
$$dh_1 = 15 \text{ µm}$$

Fig. 6.1 shows the thickness defect caused by the mill stand hysteresis with the material modulus as parameter. Therewith, decreasing the material modulus (softer steel grade material or increased temperature without modifying the composition of the material) causes increased thickness errors.

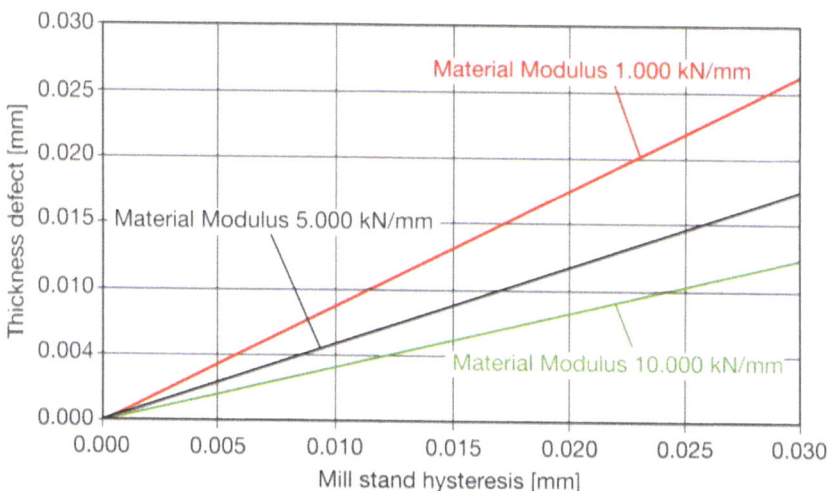

Fig. 6.1: Mill stands hysteresis and thickness defect for several material modules as parameter (Mill stands modulus 7,000 kN/mm)

Exercise 6-2: Error propagation at thickness control

In a six stand hot strip FM a transfer bar with a thickness of 44 mm is rolled to a final strip of 6.24 mm thickness. No thickness control is used. The transfer bar has a thickness defect of +1 mm. Calculate the thickness error for each FS taking into consideration the values given in **Table 6.1**.

Parameters				
Mill stand	Mill stand modulus C_G [kN/mm]	Exit thickness h_1 [mm]	Rolling force [kN]	dk_w/k_w [-]
RS		44		
F1	4,480	30.83	19,500	0.1
F2	3,990	18.72	18,000	0.08
F3	3,580	13.44	13,600	0.06
F4	3,100	10.28	12,600	0.04
F5	4,010	7.49	16,000	0.02
F6	4,080	6.24	9,900	0.01

Table 6.1: Parameters FM, part I

Register the calculated data in **Table 6.2**.

Calculated parameters (without thickness control)				
Mill stand	Absolute draft [mm]	Material modulus C_M [kN mm^{-1}]	Transfer factor Tf [-]	Exit thickness defect [mm]
F1				
F2				
F3				
F4				
F5				
F6				

Table 6.2: Parameters FM, part II

Solution:

Calculated parameters (without thickness control)				
Mill stand	Absolute draft [mm]	Material modulus C_M [kN mm^{-1}]	Transfer factor Tf [-]	Exit thickness defect [mm]
F1	13.17	740.32	0.142	0.515
F2	12.11	743.19	0.157	0.385
F3	5.28	1287.88	0.265	0.270
F4	3.16	1993.67	0.391	0.204
F5	2.79	2867.38	0.417	0.132
F6	1.25	3960.00	0.493	0.077

Table 6.2: Parameters FM, part II

The corresponding equations for calculation are:

<u>Absolute thickness draft</u>: $\Delta h = h_0 - h_1$, (6.2.1)

<u>Material modulus</u>: $C_M = \dfrac{F}{2 \cdot \Delta h}$, (6.2.2)

<u>Transfer factor</u>: $Tf = \dfrac{C_M}{C_M + C_G}$, (6.2.3)

<u>Exit thickness defect</u>:

$$dh_1 = \left(\dfrac{dk_w}{k_w} \cdot 2 \cdot \Delta h + dh_0 + ds_0 \cdot \dfrac{C_G}{C_M} \right) \cdot \dfrac{C_M}{C_M + C_G}$$
$$dh_1 = \left(\dfrac{dk_w}{k_w} \cdot 2 \cdot \Delta h + dh_0 + ds_0 \cdot \dfrac{C_G}{C_M} \right) \cdot Tf$$
(6.2.4)

In the following the above relations are derived.

Thickness errors at the exit of a rolling mill stand can occur due to

- An incoming thickness defect (Geometrical error. A strip entering the roll gap can have already a thickness defect caused by skid-marks of the reheating furnace or by uneven spraying or descaling respectively in the former stands due to changes of the strip speed.),
- An incoming defect in yield strength of the material (temperature or steel grade inhomogenities) and
- A change of the non-loaded roll gap (a wrong movement of the screw down actuators causes a geometry-related error in the strip exit thickness. Equally ranking are defects caused by wear and thermal expansion of the rolls as well as ovality of rotating parts. These kind of defects are referred to as s-defects because they origin as changes of the non-loaded roll gap. A typical example for an s-defect is the backup roll eccentricity or chatter marks in the working rolls. The origin of scatter marks is a frequency problem caused by regular changes in roll diameter in combination with critical frequencies in the complete system.

By controlled modification of screw down position these defects can be compensated. The target values are automatically calculated in the level 1 thickness control system. This necessitates certain influencing parameters according to the following equations.

The equation for roll force can be written as:

$$F = k_w \cdot b \cdot l_d = k_w \cdot b \cdot \sqrt{r \cdot \Delta h} \ . \tag{6.2.5}$$

The equation of mill stand stretch is given by

$$g = \frac{F}{C_G}. \tag{6.2.6}$$

The thickness draft Δh can be expressed as

$$\Delta h = h_0 - h_1 \tag{6.2.7}$$

and the gagemeter requation (Mill stand equation) for the exit thickness h_1 can be written as:

$$h_1 = s + \frac{F}{C_G} = s + g. \tag{6.2.8}$$

Equalizing equation (6.2.5) and (6.2.6) and inserting (6.2.7) and (6.2.8) gives

$$k_w \cdot b \cdot \sqrt{r \cdot (h_0 - h_1)} = g \cdot C_G = (h_1 - s) \cdot C_G \tag{6.2.9}$$

This equation represents in implicit manner the function

$$h_1 = f(k_w, h_0, s). \tag{6.2.10}$$

Its total differential

$$dh_1 = \frac{\partial h_1}{\partial k_w} \cdot dk_w + \frac{\partial h_1}{\partial h_0} \cdot dh_0 + \frac{\partial h_1}{\partial s} \cdot ds \tag{6.2.11}$$

provides the requested parameters for thickness control.

Derivation of equation (6.2.9) gives:

$$b \cdot k_w \cdot \frac{r \cdot (dh_0 - dh_1)}{2 \cdot \sqrt{r \cdot (h_0 - h_1)}} + b \cdot dk_w \cdot \sqrt{r \cdot (h_0 - h_1)} = (dh_0 - ds) \cdot C_G \quad (6.2.12)$$

Using (6.2.5) with $b \cdot k_w = \dfrac{F}{\sqrt{r \cdot (h_0 - h_1)}}$ for the first term

and

$\sqrt{r \cdot (h_0 - h_1)} = \dfrac{F}{b \cdot k_w}$ for the second term it can be written:

$$\frac{F}{\sqrt{r \cdot (h_0 - h_1)}} \cdot \frac{r \cdot (dh_0 - dh_1)}{2 \cdot \sqrt{r \cdot (h_0 - h_1)}} + b \cdot dk_w \cdot \frac{F}{b \cdot k_w} = (dh_1 - ds) \cdot C_G \quad (6.2.13)$$

or after rearrangement and simplifying

$$\frac{F}{2 \cdot (h_0 - h_1)} \cdot (dh_0 - dh_1) + \frac{dk_w}{k_w} \cdot F = (dh_1 - ds) \cdot C_G. \quad (6.2.14)$$

Using the material modulus $C_M = \dfrac{F}{2 \cdot (h_0 - h_1)} = \dfrac{F}{2 \cdot \Delta h}$ gives:

$$C_M \cdot (dh_0 - dh_1) + \frac{dk_w}{k_w} \cdot 2 \cdot \Delta h \cdot C_M = (dh_1 - ds) \cdot C_G$$

$$dh_0 \cdot C_M - dh_1 \cdot C_M + \frac{dk_w}{k_w} \cdot 2 \cdot \Delta h \cdot C_M = dh_1 \cdot C_G - ds \cdot C_G \quad .(6.2.15)$$

$$dh_1 = \frac{2 \cdot \Delta h \cdot C_M}{k_w \cdot (C_M + C_G)} \cdot dk_w + \frac{C_M}{C_M + C_G} \cdot dh_0 + \frac{C_G}{C_M + C_G} \cdot ds$$

The factors before the variables dk_w, dh_0 and ds are the partial differential coefficients of the function $h_1 = f(k_w, h_0, s)$. Therewith:

$$\frac{\partial h_1}{\partial k_w} = \frac{2 \cdot \Delta h \cdot C_M}{k_w \cdot (C_M + C_G)}$$
$$\frac{\partial h_1}{\partial h_0} = \frac{C_M}{C_M + C_G} \quad . \quad (6.2.16)$$
$$\frac{\partial h_1}{\partial s} = \frac{C_G}{C_M + C_G}$$

For practical application of equation (6.2.15) a different notation is requested:

$$dh_1 = \left(\frac{dk_w}{k_w} \cdot 2 \cdot \Delta h + dh_0 + ds \cdot \frac{C_G}{C_M} \right) \cdot \frac{C_M}{C_M + C_G} \quad (6.2.17)$$

at which the differentials are substituted by the finite defect parameters.

$\dfrac{dk_w}{k_w}$ in equation (6.2.17) correlates with the relative error of the material flow stress k_w.

Differentiation of equation (6.2.17) according to time results in the defect transmission speed:

$$\frac{dh_1}{dt} = \left(\frac{\frac{dk_w}{k_w}}{dt} \cdot 2 \cdot \Delta h + \frac{dh_0}{dt} + \frac{ds}{dt} \cdot \frac{C_G}{C_M} \right) \cdot \frac{C_M}{C_M + C_G} . \qquad (6.2.18)$$

In (6.2.17) and (6.2.18) defects in material flow stress, strip geometry and s-defects are treated as comparable parameters:

- Flow stress defect: $\dfrac{dk_w}{k_w} \cdot 2 \cdot \Delta h$ resp. $\dfrac{\frac{dk_w}{k_w}}{dt} \cdot 2 \cdot \Delta h$,

- Geometrical entry defect: dh_0 resp. $\dfrac{dh_0}{dt}$,

- s-entry defect: $ds \cdot \dfrac{C_G}{C_M}$ resp. $\dfrac{ds}{dt} \cdot \dfrac{C_G}{C_M}$.

An entry defect - independent from its origin - is transmitted with the transfer factor

$$Tf = \frac{C_M}{C_M + C_G} \qquad (6.2.19)$$

onto the mill stands exit side.

In case the exit thickness defect $dh_1 = 0$ is demanded equations (6.2.17) and (6.2.18) provide the target values for the mill screw down position:

$$ds = -\left(\frac{dk_w}{k_w} \cdot 2 \cdot \Delta h + dh_0\right) \cdot \frac{C_M}{C_G} \qquad (6.2.20)$$

and

$$\frac{ds}{dt} = -\left(\frac{\frac{dk_w}{k_w}}{dt} \cdot 2 \cdot \Delta h + \frac{dh_0}{dt}\right) \cdot \frac{C_M}{C_G}. \qquad (6.2.21)$$

Equation (6.2.20) gives the change of screw down position for the compensation of material flow stress defects and/or geometrical entry defects. Formula (6.2.21) provides the respective screw down speeds for the defect compensation.

The material modulus correlates with the partial differential of the function $F = f(\Delta h)$ and characterizes the change in roll force if the absolute draft in the pass is modified. This means:

$$C_M = \frac{\partial F}{\partial \Delta h}. \qquad (6.2.22)$$

The equation for roll force is:

$$F = k_w \cdot b \cdot l_d = k_w \cdot b \cdot \sqrt{r \cdot \Delta h}. \qquad (6.2.5)$$

For the partial differential follows:

$$\frac{\partial F}{\partial \Delta h} = b \cdot k_w \cdot \frac{r}{2 \cdot \sqrt{r \cdot \Delta h}} \qquad (6.2.23)$$

and using $b \cdot k_w = \dfrac{F}{\sqrt{r \cdot \Delta h}}$ it can be written

$$\frac{\partial F}{\partial \Delta h} = \frac{F}{\sqrt{r \cdot \Delta h}} \cdot \frac{r}{2 \cdot \sqrt{r \cdot \Delta h}} = \frac{F}{2 \cdot \Delta h} \qquad (6.2.24)$$

and therewith

$$C_M = \frac{\partial F}{\partial \Delta h} = \frac{F}{2 \cdot \Delta h}. \qquad (6.2.25)$$

For a rigid mill stand with its modulus very large compared to the material modulus $\left(\dfrac{C_M}{C_G} \ll 1\right)$ the transmitted thickness defect follows according to equation (6.2.17)

$$dh_1 \approx ds \qquad (6.2.26)$$

and for anon rigid mill stand with its modulus very small compared to the material modulus $\left(\dfrac{C_M}{C_G} \gg 1\right)$

$$dh_1 \approx \frac{dk_w}{k_w} \cdot 2 \cdot \Delta h + dh_0. \qquad (6.2.27)$$

This means for a rigid mill stand mainly changes in the non-loaded roll gap (s-defects) are relevant and for a non rigid mill stand geometry and flow stress defects are transferred to the mill stand exit. An analogue situation is given according to (6.2.18) for the defect transmission speed.

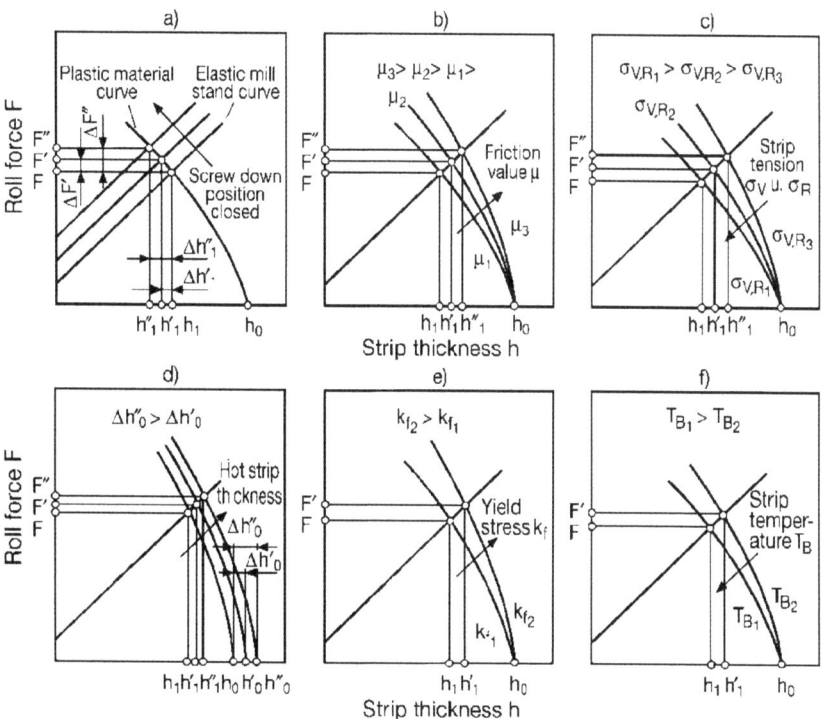

Fig. 6.2: Roll force-strip thickness patterns /9/

Summarizing, the following can be noted, **Fig. 6.2**: Roll force and thickness reduction increase when closing screw down position. If the screw downs are opened the opposite effect occurs, Fig. 6.2a. If the screw downs remain constant, i.e. by given elastic mill stretch curve, roll force increases and thickness draft reduces if friction value increases and strip tension decreases, Fig. 6.2b and 6.2c. In the same context roll force increases and thickness errors occur on the exit strip if variations take place on the entry strip thickness, Fig. 6.2d. Variations in yield strength of the hot strip, caused by different chemical composition or by variation in strip temperature (e.g. skid-mark), result in thickness defects in the exit strip, Fig. 6.2e and 6.2f.

7. Measuring techniques in HSM's

Exercise 7-1: Width measurement

A HSM is equipped with a strip width measurement system at the exit of the FM. The system operates with back light and one CCD-camera. Determine the absolute and relative measuring error for a horizontal "flying strip" with a width of 1500 mm and a wave height of 100 mm. The CCD-camera is mounted 2000 mm above the roller table.

Solution:

The solution is given according to the intercept theorem, **Fig. 7.1**:

$$\frac{d_{Strip-Camera}}{d_{Rollertable-Camera}} = \frac{b'}{b} \text{ noting } d_{Strip-Camera} = 1,900 \text{ mm}$$

$$d_{Rollertable-Camera} = 2,000 \text{ mm}$$

and the strip width $b = 1,500$ mm using

$$b' = \frac{d_{Strip-Camera}}{d_{Rollertable-Camera}} \cdot b = \frac{1,900 \text{ mm}}{2,000 \text{ mm}} \cdot 1,500 \text{ mm} = 1,425 \text{ mm}.$$

The absolute measuring error is 75 mm and the relative measuring error 5%.

Fig. 7.2 to **Fig. 7.4** show the absolute and relative measuring error for several strip widths and the two assumed wave heights of 100 mm and 50 mm. The relative measuring error is independent from the strip width.

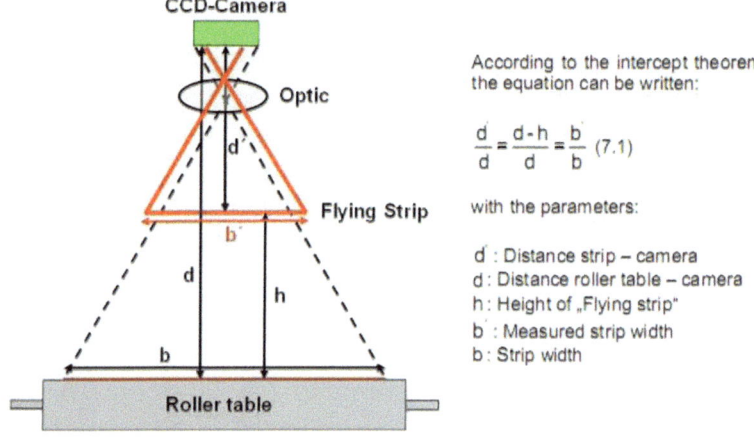

According to the intercept theorem the equation can be written:

$$\frac{d'}{d} = \frac{d-h}{d} = \frac{b'}{b} \quad (7.1)$$

with the parameters:

d' : Distance strip – camera
d : Distance roller table – camera
h : Height of „Flying strip"
b' : Measured strip width
b : Strip width

Fig. 7.1: Pattern of width measurement of hot strip using a CCD-camera

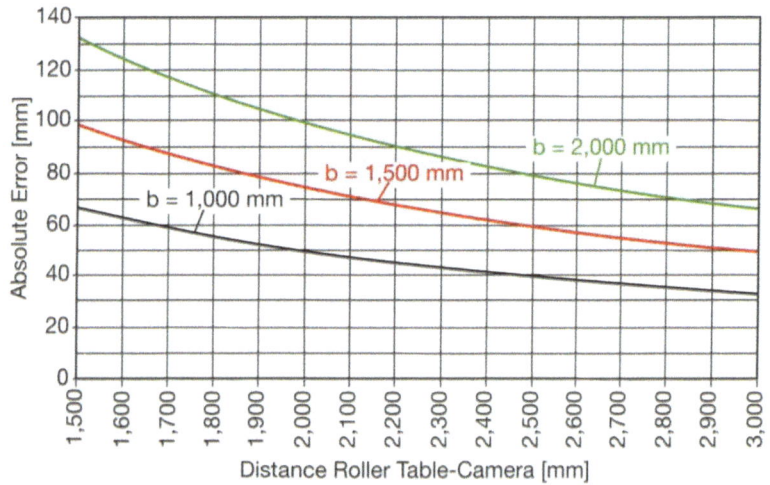

Fig. 7.2: Phenomena „Flying Strip". Absolute measuring error for different strip widths and a wave height h = 100 mm

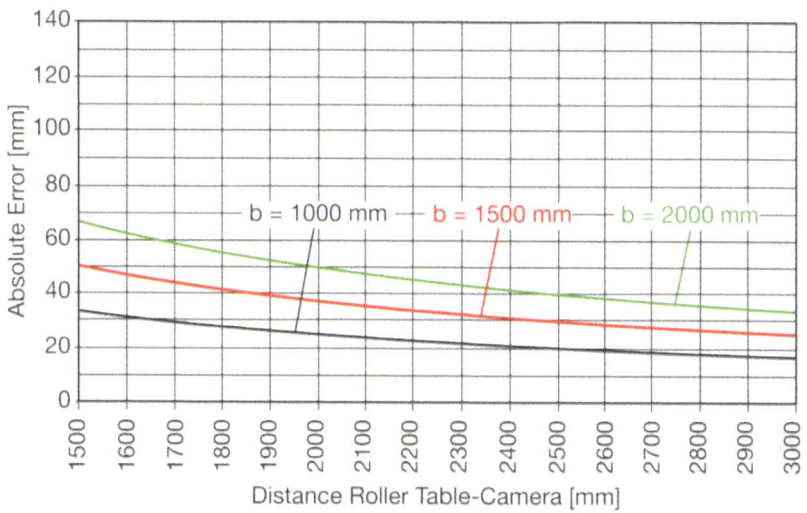

Fig. 7.3: Phenomena „Flying Strip". Absolute measuring error for different strip widths and a wave height h = 50 mm

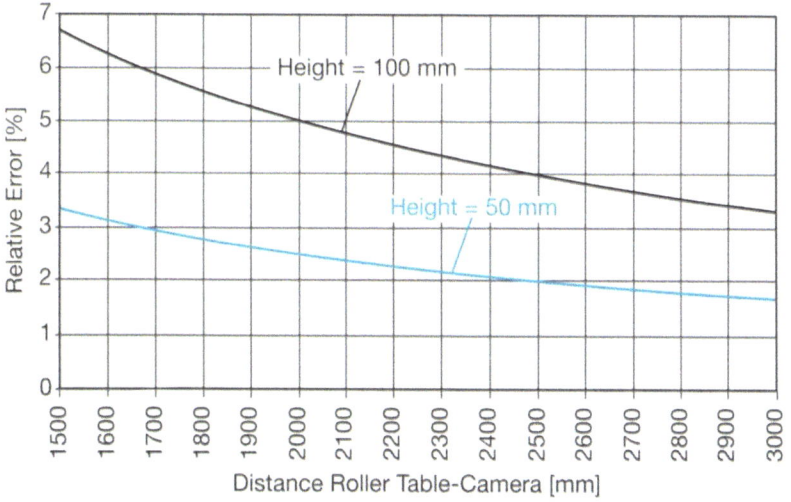

Fig. 7.4: Phenomena „Flying Strip ".Relative measuring error for the wave heights h = 100 mm and h = 50 mm

Remark

For avoiding measuring errors of strip width due to "flying" or angled moving strips in the measuring field stereoscopic measuring systems are applied in HSM's. This technique allows further registering the strip position and strip camber within the HSM.

Investigations on several HSM's have shown the dependency of thickness profile wedge on strip tracking and strip position within the rolling mill. For rolling of a hot strip with a homogeneous wedge-free thickness profile the lateral movement of the strip in the roll gap has to be minimized.

Several actions can be undertaken in the HSM for minimizing lateral strip movement:

- Implementation of heavy hydraulically operating side guides in the RM and FM.
- Tilting of screw down actuators in the RM and FM.

Reason for camber formation at the transfer bar can be:

- Non homogeneous heated slabs in the reheating furnaces resulting in different yield stresses within the slab width when rolled in the RM.
- Length cut slabs from the caster with wedged thickness profile.
- Different mill stand stretch behavior on the operator and driven side in the RS.
- Non parallel roll gap in RM due to wrong work roll grind from the roll shop or wrong roll gap calibration.

- Missing or wrong calibration of the entry side guides of the RS causing sloping slab tracking behavior in the roll gap.
- Differences in friction over width (roughness, water).

Exercise 7-2: Thickness measurement

In this exercise the physical equations for determination of the possible thickness gauging errors due to several ambient influences are treated. Therefore as an introduction the relevant formula work is derived /32/. In HSM's isotopic or X-ray strip irradiating techniques are commonly used as strip thickness gauges.

A. Constants for consideration the measuring errors

Constant	Formula/Numerical value/Comment	Unit
Attenuation coefficient μ	$\mu = \mu' \cdot \rho$ with the quantities mass absorption coefficient μ' and material density ρ	cm^{-1}
Mass absorption coefficient μ'	Iron: $\mu'_{Fe} = 0.075$; Oxygen: $\mu'_{O} = 0.08$; Hydrogen: $\mu'_{H} = 0.08$; Air: $\mu'_{Air} = 0.08$ For approximation: $\mu'_{Fe} = \mu'_{O} = \mu'_{H} = \mu'_{Air} = 0.08$	cm^2/g
Material density ρ	Iron: $\rho_{Fe} = 7.86$; Scale: $\rho_{Scale} = 2.5 - 5$ Water: $\rho_{Water} = 1$	g/cm^3
Mass absorption coefficient μ' for chemical compounds	$\mu' = \dfrac{\sum_n \mu'_n \cdot (A \cdot W)_n}{\sum_n (A \cdot W)_n}$ A: Atomic weight of the chemical element in the compound W: Valence of the chemical element in the compound For approximation: $\mu'_{Fe} = \mu'_{O} = \mu'_{H} = \mu'_{Air} = 0.08$	cm^2/g

Table 7.1: Constants for calculation of the measuring errors

B. Derivation of the mathematical equations

B1. Geometry of the measuring apparatus (Shift of measuring distance)

Thermal expansion and shrinkage of the measuring house or frame gives a shift of the measuring distance D_{Air} by ΔD_{Air} and therefore a shift of registered radiation intensity I according to:

$$I_0 \sim \frac{1}{(D_{Air})^2}, \qquad (B1.1)$$

(Quadratic distance law without shift of measuring distance).

$$I_1 \sim \frac{1}{(D_{Air} + \Delta D_{Air})^2} \qquad (B1.2)$$

(Quadratic distance law with shift of measuring distance).

The ratio $\dfrac{\Delta I}{I_0}$ follows according to the relationship:

$$\frac{\Delta I}{I_0} = \frac{I_1 - I_0}{I_0} = \frac{I_1}{I_0} - 1 = \frac{D_{Air}^2}{(D_{Air} + \Delta D_{Air})^2} - 1$$

$$\frac{\Delta I}{I_0} = \left(\frac{D_{Air}}{D_{Air} + \Delta D_{Air}}\right)^2 - 1 \qquad (B1.3)$$

The thickness gauging error is calculated using the attenuation law with the radiation intensity I_0^* without absorber according to:

$$I_0 = I_0^* \cdot \exp(-\mu_{Fe} \cdot D_{Fe}), \tag{B1.4}$$

(Attenuation law for a strip with thickness D_{Fe} without shift of measuring distance)

$$I_1 = I_0^* \cdot \exp\left[-\mu_{Fe} \cdot (D_{Fe} + \Delta D_{Fe})\right]. \tag{B1.5}$$

(Attenuation law for a strip with an apparent thickness $D_{Fe} + \Delta D_{Fe}$ caused by distance shift)

This gives:

$$\frac{\Delta I}{I_0} = \frac{I_1 - I_0}{I_0} = \frac{I_1}{I_0} - 1 = \frac{I_0^* \cdot \exp\left[-\mu_{Fe} \cdot (D_{Fe} + \Delta D_{Fe})\right]}{I_0^* \cdot \exp(-\mu_{Fe} \cdot D_{Fe})} - 1$$

$$\frac{\Delta I}{I_0} = \exp(-\mu_{Fe} \cdot \Delta D_{Fe}) - 1 \tag{B1.6}$$

Equalizing (B1.3) and (B1.4) gives:

$$\left(\frac{D_{Air}}{D_{Air} + \Delta D_{Air}}\right)^2 - 1 = \exp(-\mu_{Fe} \cdot \Delta D_{Fe}) - 1. \tag{B1.7}$$

Simplifying finally gives for the measuring error ΔD_{Fe}:

$$\Delta D_{Fe} = \frac{-2 \cdot \ln\left(\dfrac{D_{Air}}{D_{Air} + \Delta D_{Air}}\right)}{\mu_{Fe}}. \tag{B1.8}$$

B2. Sloping strip angles (Flying strip head ends)

In case the radiation source irradiates the strip with a slope (sloping angle α) the thickness measuring error ΔD_{Fe} is calculated according to the cosine of the sloping angle, **Fig. B2.1**.

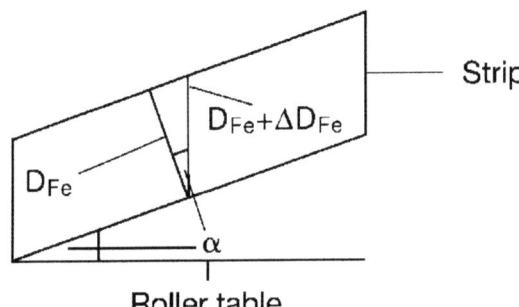

Fig. B2.1: Geometric situation in case of transversal irradiated strip

It is true:

$$\cos\alpha = \frac{D_{Fe}}{D_{Fe} + \Delta D_{Fe}}. \qquad (B2.1)$$

The absolute measuring error is according to

$$\Delta D_{Fe} = \frac{D_{Fe}}{\cos\alpha} - D_{Fe} = D_{Fe} \cdot \left(\frac{1}{\cos\alpha} - 1\right) \qquad (B2.2)$$

and the relative error

$$\frac{\Delta D_{Fe}}{D_{Fe}} = \frac{1}{\cos\alpha} - 1. \qquad (B2.3)$$

B3. Water on the radiation source protection or on the strip surface

A water film with a thickness ΔD_{Water} on the radiation source protection or on the strip surface affects strip thickness gauging as follows:

$$I_0 = I_0^* \cdot \exp(-\mu_{Fe} \cdot D_{Fe}), \qquad (B3.1)$$

(Attenuation law for a strip with a thickness of D_{Fe} without water disturbances)

$$I_1 = I_0^* \cdot \exp(-\mu_{Fe} \cdot D_{Fe}) \cdot \exp(-\mu_{Water} \cdot \Delta D_{Water}), \qquad (B3.2)$$

(Attenuation law for a strip with a thickness of D_{Fe} and an additional water film)

The resulting thickness gauging results according to the equation:

$$\begin{aligned} I_1 &= I_0^* \cdot \exp(-\mu_{Fe} \cdot D_{Fe}) \cdot \exp(-\mu_{Water} \cdot \Delta D_{Water}) \\ I_1 &= I_0^* \cdot \exp\left[-\mu_{Fe} \cdot (D_{Fe} + \Delta D_{Fe})\right] \end{aligned} \qquad (B3.3)$$

Simplifying gives

$$\Delta D_{Fe} = \frac{\mu_{Water}}{\mu_{Fe}} \cdot \Delta D_{Water}. \qquad (B3.4)$$

Introducing the mass absorption coefficient $\mu' = \mu \cdot \rho$ in above equation leads to

$$\Delta D_{Fe} = \frac{\mu'_{Water}}{\mu'_{Fe}} \cdot \frac{\rho_{Water}}{\rho_{Fe}} \cdot \Delta D_{Water} \qquad (B3.5)$$

and together with $\mu'_{Wasser} \approx \mu'_{Fe}$ finally to

$$\Delta D_{Fe} = \frac{\rho_{Water}}{\rho_{Fe}} \cdot \Delta D_{Water}. \qquad (B3.6)$$

B4. Scale on radiation source protection or on strip surface

The derivation of the resulting gauging error due to scale of thickness ΔD_{Scale} can be performed in analogy to chapter B3. Together with the approach $\mu'_{Scale} \approx \mu'_{Fe}$ it follows

$$\Delta D_{Fe} = \frac{\rho_{Scale}}{\rho_{Fe}} \cdot \Delta D_{Scale}. \qquad (B4.1)$$

Scale for rolling temperatures between 800°C and 900°C at the exit of the finishing mill exists to a wustite (FeO) component between 90% and 95%. Therefore the approach $\mu'_{Scale} \approx \mu'_{Fe}$ is valid.

B5. Shift of ambient air temperature

Temperature shift of the irradiated air column affects thickness measuring accuracy according to:

$$I_0 = I_0^* \cdot \exp(-\mu_{Fe} \cdot D_{Fe}) \cdot \exp(-\mu_{0Air} \cdot D_{Air})$$
$$I_0 = I_0^* \cdot \exp(-\mu_{Fe} \cdot D_{Fe}) \cdot \exp(-\mu'_{Air} \cdot \rho_{Air}(T_0) \cdot D_{Air})$$
(B5.1)

(Attenuation law for a strip with a thickness D_{Fe} and an irradiated air column with temperature $T=T_0$)

$$I_0 = I_0^* \cdot \exp(-\mu_{Fe} \cdot D_{Fe}) \cdot \exp(-\mu_{1Air} \cdot D_{Air})$$
$$I_0 = I_0^* \cdot \exp(-\mu_{Fe} \cdot D_{Fe}) \cdot \exp(-\mu'_{Air} \cdot \rho_{Air}(T_1) \cdot D_{Air})$$
(B5.2)

(Attenuation law for a strip with a thickness D_{Fe} and an irradiated air column with temperature $T=T_1$)

The total thickness measuring error follows according to the difference of the individual errors $\Delta D_{Fe,0}$ and $\Delta D_{Fe,1}$ calculated for the temperatures T_0 and T_1 of the irradiated air column. The individual errors $\Delta D_{Fe,0}$ and $\Delta D_{Fe,1}$ follow according to the equations:

$$I_0 = I_0^* \cdot \exp(-\mu_{Fe} \cdot D_{Fe}) \cdot \exp(-\mu'_{Air} \cdot \rho_{Air}(T_0) \cdot D_{Air})$$
$$I_0 = I_0^* \cdot \exp\left[-\mu_{Fe} \cdot (D_{Fe} + \Delta D_{Fe,0})\right]$$
(B5.3)

$$I_1 = I_0^* \cdot \exp(-\mu_{Fe} \cdot D_{Fe}) \cdot \exp(-\mu'_{Air} \cdot \rho_{Air}(T_1) \cdot D_{Air})$$
$$I_1 = I_0^* \cdot \exp\left[-\mu_{Fe} \cdot (D_{Fe} + \Delta D_{Fe,1})\right]$$
(B5.4)

Using the approach $\mu'_{Air} \approx \mu'_{Fe}$ and simplifying, the individual errors are given by

$$\Delta D_{Fe,0} = \frac{\rho_{Air}(T_0)}{\rho_{Fe}} \cdot \Delta D_{Air}$$
(B5.5)

and

$$\Delta D_{Fe,0} = \frac{\rho_{Air}(T_1)}{\rho_{Fe}} \cdot \Delta D_{Air}.$$ (B5.6)

Temperature shift of the air column from T_0 to T_1 gives

$$\Delta D_{Fe} = \Delta D_{Fe,1} - \Delta D_{Fe,0} = \frac{\rho_{Air}(T_1) - \rho_{Air}(T_0)}{\rho_{Fe}} \cdot D_{Air}.$$ (B5.7)

as thickness measuring error.

For $T_1 > T_0$ there is $\rho_{Air}(T_1) < \rho_{Air}(T_0)$ and therefore $\Delta D_{Fe} < 0$ (negative temperature dependency of measuring error).

B6. Water steam

Shifting temperature of the water steam column during rolling gives a thickness error in analogy to the influence "Shift of ambient air temperature", which is already described in chapter B5. The total thickness measuring error results from the difference of the individual errors $\Delta D_{Fe,0}$ and $\Delta D_{Fe,1}$ arising at the temperatures T_0 and T_1 of the irradiated water steam column $D_{Watersteam}$. The individual errors can be written as

$$\Delta D_{Fe,0} = \frac{\mu'_{Watersteam}}{\mu'_{Fe}} \cdot \frac{\rho_{Watersteam}(T_0)}{\rho_{Fe}} \cdot D_{Watersteam}$$

$$\Delta D_{Fe,0} = \frac{\rho_{Watersteam}(T_0)}{\rho_{Fe}} \cdot D_{Watersteam}$$ (B6.1)

and

$$\Delta D_{Fe,1} = \frac{\mu'_{Watersteam}}{\mu'_{Fe}} \cdot \frac{\rho_{Watersteam}(T_1)}{\rho_{Fe}} \cdot D_{Watersteam}$$

$$\Delta D_{Fe,1} = \frac{\rho_{Watersteam}(T_1)}{\rho_{Fe}} \cdot D_{Watersteam}$$ (B6.2)

In both equations $\mu'_{Watersteam} = \mu'_{Fe}$ is assumed.

Therefore the resulting measuring error is given according to

$$\Delta D_{Fe} = \Delta D_{Fe,1} - \Delta D_{Fe,0}$$

$$\Delta D_{Fe} = \frac{\rho_{Watersteam}(T_1) - \rho_{Watersteam}(T_0)}{\rho_{Fe}} \cdot D_{Watersteam}$$ (B6.3)

For $T_1 > T_0$ there is $\rho_{Watersteam}(T_1) > \rho_{Watersteam}(T_0)$ and therefore $\Delta D_{Fe} > 0$ (positive temperature dependency of measuring error).

Sudden occurrence of water steam causes an error in analogy of chapter B3 "Water on radiation source protection or on strip surface". The resulting measuring error for sudden occurrence of water steam therefore is given by

$$\Delta D_{Fe} = \frac{\rho_{Watersteam}(T)}{\rho_{Fe}} \cdot D_{Watersteam} \cdot$$ (B6.4)

This equation assumes $\mu'_{Watersteam} = \mu'_{Fe}$.

B7. Temperature shift in strip (No modeling of temperature compensation)

Temperature shift in the strip in case of missing or wrong temperature compensation cause an error due to:

$$I_0 = I_0^* \cdot \exp(-\mu_{Fe,0} \cdot D_{Fe}) = I_0^* \cdot \exp(-\mu'_{Fe} \cdot \rho_{Fe}(T_0) \cdot D_{Fe}). \quad (B7.1)$$

(Attenuation law for a strip with a thickness D_{Fe} and a temperature $T=T_0$)

$$I_1 = I_0^* \cdot \exp(-\mu_{Fe,1} \cdot D_{Fe}) = I_0^* \cdot \exp(-\mu'_{Fe} \cdot \rho_{Fe}(T_1) \cdot D_{Fe}). \quad (B7.2)$$

(Attenuation law for a strip with a thickness D_{Fe} and a temperature $T=T_1$; the temperature shift can be caused e.g. by skid marks from the furnaces or a hot rolled strip head end)

The thickness measuring error due to strip temperature shift from T_0 to T_1 the following equation can be derived:

$$\begin{aligned} I_1 &= I_0^* \cdot \exp\left[-\mu'_{Fe} \cdot \rho_{Fe}(T_0) \cdot (D_{Fe} + \Delta D_{Fe})\right] \\ I_1 &= I_0^* \cdot \exp(-\mu'_{Fe} \cdot \rho_{Fe}(T_1) \cdot D_{Fe}) \end{aligned} \quad (B7.3)$$

By simplifying this equation it follows

$$\Delta D_{Fe} = \left(\frac{\rho_{Fe}(T_1)}{\rho_{Fe}(T_0)} - 1\right) \cdot D_{Fe} \quad (B7.4)$$

for the absolute and

$$\frac{\Delta D_{Fe}}{D_{Fe}} = \frac{\rho_{Fe}(T_1)}{\rho_{Fe}(T_0)} - 1 \qquad (B7.5)$$

for the relative measuring error.

Fig. B7.1 gives the essential correction of thickness measurement regarding shift of hot rolled stock temperatures (Dilatometer curves).

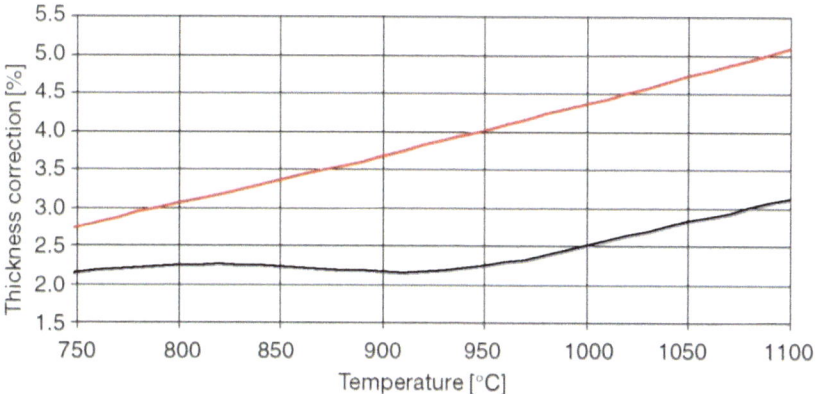

Fig. B7.1: Dilatometer curves showing the influence of rolled stock temperature on accuracy of thickness measurement. (Red curve: Fe containing Mg; black curve: Fe containing C < 0.09%), Source: IMS.

Exercise 7-2a:

Determine the relative thickness measuring error for a low carbon steel with a rolling temperature of 900°C in case of missing temperature compensation in the gauging device.

Solution:

According to the dilatometer curves in Fig. B7.1 the relative measuring error is roughly 2.2%.

Exercise 7-2b

Calculate the absolute thickness measuring error for a hot strip with a final thickness of 2 mm and a thermal expansion of the measuring house of 2 mm. The basic distance between radiation source and detector at room temperature is 2000 mm. The attenuation coefficient for iron is $\mu_{Fe} = 5.895 \, cm^{-1}$.

Solution:

The absolute thickness measuring error ΔD_{Fe} can be determined according to equation (B1.8)

$$\Delta D_{Fe} = \frac{-2 \cdot \ln\left(\dfrac{D_{Luft}}{D_{Luft} + \Delta D_{Luft}}\right)}{\mu_{Fe}}. \tag{B1.8}$$

Inserting the given data from the exercise gives

$$\Delta D_{Fe} = \frac{-2 \cdot \ln\left(\frac{2000 \text{ mm}}{2000 \text{ mm} + 2 \text{ mm}}\right)}{58.95 \text{ mm}^{-1}}.$$

$\Delta D_{Fe} = 34 \text{ µm}$

Exercise 7-2c

Determine the absolute thickness measuring error for a strip with a final thickness of 2 mm moving with a sloping angle of 15° under the thickness gauge.

Solution:

In case the radiation source irradiates the strip with a slope (sloping angle α) the thickness measuring error ΔD_{Fe} is calculated according to the cosine of the sloping angle, Fig. B2.1.

The absolute thickness measuring error is determined according to

$$\Delta D_{Fe} = \frac{D_{Fe}}{\cos \alpha} - D_{Fe} = D_{Fe} \cdot \left(\frac{1}{\cos \alpha} - 1\right). \tag{B2.2}$$

Inserting the given values according to exercise provides $\Delta D_{Fe} = 71 \text{ µm}$.

Exercise 7-2d

Determine the arising absolute thickness measuring error for a strip with a final thickness of 2 mm covering a 2 mm thick water film on its surface in the area of thickness gauge.

Solution:

The relevant equation for calculating the measuring error is:

$$\Delta D_{Fe} = \frac{\rho_{Water}}{\rho_{Fe}} \cdot \Delta D_{Water} \tag{B3.6}$$

Using $\rho_{Water} = 1\, t\, m^{-3}$

and $\rho_{Fe} = 7.86\, t\, m^{-3}$ gives a measuring error of $\Delta D_{Fe} = 127\, \mu m$.

Exercise 7-2e

Calculate the arising absolute thickness measuring error for a strip with a final thickness of 2 mm in case of sudden temperature shift of the irradiated air column from 20°C to 100°C immediately after pass begin. The density of air is
$\rho_{Air}(T = 20\,°C) = 1.205 \cdot 10^{-3}\, \frac{g}{cm^3}$,

$\rho_{Air}(T = 100\,°C) = 0.946 \cdot 10^{-3}\, \frac{g}{cm^3}$.

The distance between radiation source and radiation detector is assumed with 2 m.

Solution:

Temperature shift of the air column from T_0 to T_1 gives

$$\Delta D_{Fe} = \Delta D_{Fe,1} - \Delta D_{Fe,0} = \frac{\rho_{Air}(T_1) - \rho_{Air}(T_0)}{\rho_{Fe}} \cdot D_{Air} . \qquad (B5.7)$$

as thickness measuring error.

For $T_1 > T_0$ there is $\rho_{Air}(T_1) < \rho_{Air}(T_0)$ and therefore $\Delta D_{Fe} < 0$ (negative temperature dependency of measuring error).

Evaluation with the given values according to exercise provides $\Delta D_{Fe} = -66\ \mu m$ as absolute measuring error.

Exercise 7-2f

Which practical measures should be undertaken for minimizing the ambient influences on the accuracy of thickness gauging? Distinguish in the numbering between the influencing parameters

- Geometry of the thickness gauging apparatus,
- Sloping strip moving angles,
- Water and water steam in the measuring field and
- Shift of the ambient air temperature.

Solution:

Geometry of the thickness gauging apparatus (Shift of distance between radiation source and detector unit).Due to the alternating thermal expansion or thermal shrinking of the measuring house or measuring frame an alternating variation of the distance between radiation source and detector unit takes place. This angle dependent movement can be registered by installing inclinometers at the measuring apparatus. The inclinometer measuring can be used to compensate this effect on thickness measurement. Another possibility for preventing this effect is to use thermal resistant material for the measuring house or the frame for isolation against heat. The measuring frame e.g. can be thermally isolated by an internal water cooling circuit at constant temperatures of 20°C.

For the elimination of thermal long time effects the thickness measuring gauge should be calibrated several times during rolling breaks.

Sloping strip moving angles ("Flying strip head ends"). Implementing a strip height distribution measuring system, e.g. TopPlan system provided by IMS Messtechnik in Germany, enables the determination of all irradiation angles for the strip area in the thickness measuring area. Compensation of thickness signal due to some trigonometric functions ispossible very easily.

Another practice is to block thickness monitor control until the down coiler has gripped the incoming strip head end if a parametrical limit of the strip height is exceeded.

The best practice for minimization the "jumping effect" of the strip head end is the mechanical adaption of the pass lines rolling mill and run out table on the same height level by using triangle shaped retractable bars, which are put below the upper work roll sets.

Water and water steam. Principally the installation of a light curtain in the thickness measuring area and measuring the different intensities of the back scattered light in the case of occurring water steam is possible. The principle bases on Rayleigh-scattering, which gives the scattering effect of the especially short wavelength blue violet in dependence on water steam density. Providing a calibrated sensor the density of the water steam can be extracted and used for compensation of the thickness measurement.

A more easy and practicable procedure is the installation of large fan blowers and cross sprayings before and after the thickness measurement to guarantee a water and water steam free measuring field.

Further the implementation of internally cooled roller table rolls is recommended.

To minimize vanishing water from the work roll cooling system onto the strip the condition of the roll wipers should be checked and regularly maintained.

From theory it can be deduced to reduce the distance between radiation source and detector unit for minimizing the ambient influence of water steam on the thickness measurement accuracy.

Ambient air temperature. The registration of ambient air temperature is principally possible by installation a small pitot static tube for the integration of fast responding thermocouples measuring the air temperature. Practical tests at a hot strip mill had shown some problems occur due to intake of turbulence free surrounding air in the tube. Furthermore such a system needs much maintenance effort due to its essential calibration procedure.

Regarding theory follows similar to the parameter water steam that the influence of ambient air on thickness measuring accuracy depends on the distance between radiation source and detector unit. For decreasing the influence of ambient air temperature on thickness measuring accuracy the distance between radiation source and detector has to be shortened.

8. Roll flattening and minimum rollable strip thickness

The relevant equations for solving the exercises are given in Chapter 7 of the book "Basics in flat rolling" /27/.

Exercise 8-1: Specific roll force and roll flattening

A strip is rolled in a CRM with a specific roll force of $\frac{F}{b} = 10 \ \frac{kN}{mm}$. The flattening constant of the steel rolls is $C_H = 22 \cdot 10^{-6} \ mm^2 \ N^{-1}$. Calculate the absolute thickness draft Δh for the limit case $\frac{r'}{r} = 2$.

Solution:

According to Hitchcock the ratio between the flattened and non flattened roll radius is given by:

$$\frac{r'}{r} = \left(1 + C_H \cdot \frac{F}{b \cdot \Delta h}\right). \qquad (8.7)$$

Reorganization and introduction of the given values provides:

$$\Delta h = C_H \cdot \frac{F}{b \cdot \left(\frac{r'}{r} - 1\right)}$$

$$\Delta h = 22 \cdot 10^{-6} \; \frac{mm^2}{N} \cdot \frac{10 \cdot 10^3}{1 \cdot (2-1)} \; \frac{N}{mm}. \tag{8.8}$$

$\Delta h = 0.22 \; mm$

Fig. 8.2 show the relationship between absolute thickness draft and the ratio $\frac{r'}{r}$. The smaller the specific roll force the smaller is the resulting radii ratio between the flattened and non-flattened roll. Absolute thickness drafts of more than one millimeter are typical for HSM's; therefore the influence of roll flattening phenomena on roll force and torque can be neglected in first approximations in HSM's. In CRM's with thickness drafts of less than one millimeter the influence of roll flattening on roll force and roll torque cannot be neglected; the radii ratio $\frac{r'}{r}$ is shifted towards higher values for increasing specific roll forces and larger thickness drafts.

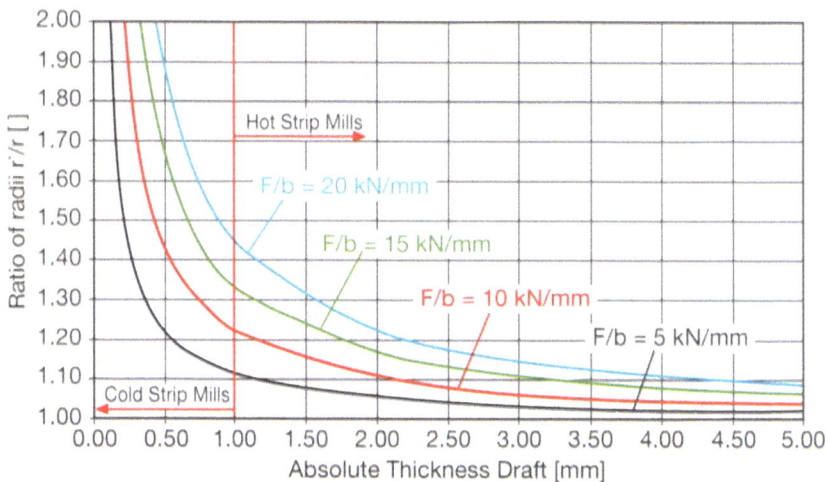

Fig. 8.2: Absolute thickness draft and roll flattening for several specific roll forces

Exercise 8-2: Minimum roll able strip thickness and roll radius

Calculate the minimum roll able strip thickness for the steel grade S235JR with the following process parameters: Work roll radius $r = 350$ mm, sliding friction coefficient $\mu = 0.1$, average yield stress $k_{fm} = 454$ N mm^{-2} and flattening constant for steel rolls $C_H = 22 \cdot 10^{-6}$ mm^2 N^{-1}.

Solution:

The Hitchcock equation for calculating the minimum rollable strip thickness in case of sliding friction is:

$$h_{min} = 1.545 \cdot \mu \cdot C_H \cdot k_f \cdot r. \tag{8.4}$$

Inserting the given rolling parameters provides:

$$h_{min} = 1.545 \cdot 0.1 \cdot 22 \cdot 10^{-6} \text{ mm}^2 \text{ N}^{-1} \cdot 454 \text{ N mm}^{-2} \cdot 350 \text{ mm}$$
$$h_{min} = 0.54 \text{ mm}$$
(8.9)

There are several possibilities to reduce the minimum rollable strip thickness:

(1) Reduction of the friction value in the roll gap using appropriate rolling oils and emulsions,

(2) Reduction of work roll radius as in the case of MRS (Work roll radius for 20-high mill stand: 30 mm) and

(3) Application of work rolls with a higher elasticity modulus, such as ceramic rolls.

Fig. 8.3 shows for several friction coefficients the minimum rollable strip thickness in dependency on the work roll radius. The considered steel grade is S235JR and the roll grade is cast steel. With a work roll radius of 50 mm rolling of foil material with a thickness of some 10 µm is possible on a 20-high mill stand configuration.

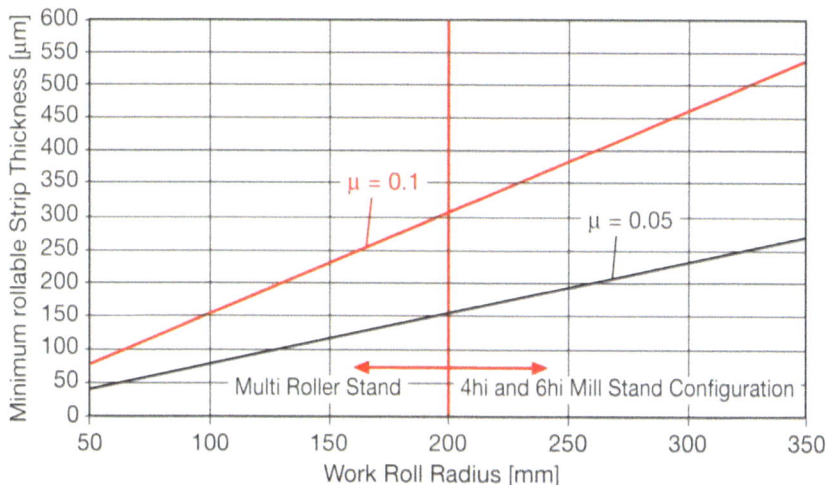

Fig. 8.3: Minimum rollable strip thickness and roll radius for different friction coefficients

9. Operational data of HSM's

An extended description and definition of operational data in HSM's is given in Chapter 2 of the book "Modern Hot Strip Production" /32/.

Exercise 9-1: Time and production figures of a hot wide strip mill

A semi-continuously operating hot wide strip mill produces 563 t h^{-1} in the net rolling time. Determine the yearly mill production taking into account the following boundary conditions:

- Other idle times $t_{IT} = 47$ h a^{-1},
- Planned maintenance times $t_{MT} = 902$ h a^{-1},
- Roll changing times $t_{RCT} = 533$ h a^{-1}.
- Failure times of 14.01% of gross rolling time.

Solution:

For calculation of the yearly production from the given data the knowledge of gross t_{GRT} and net rolling time t_{NRT} is necessary. Gross rolling time follows from calendar time t_{CT} subtracting the other idle times t_{IT} and times for planned mill repairs and maintenance t_{MT}:

$$t_{GRT} = t_{CT} - t_{IT} - t_{MT}$$
$$t_{GRT} = 8,760 \text{ h a}^{-1} - 47 \text{ h a}^{-1} - 902 \text{ h a}^{-1} = 7,811 \text{ h a}^{-1}. \quad (9.1.1)$$

The failure time t_{FT} is:

$$t_{FT} = 0.1401 \cdot t_{GRT}$$
$$t_{FT} = 0.1401 \cdot 7{,}811 \, h \, a^{-1} = 1{,}094 \, h \, a^{-1}.$$
(9.1.2)

The net rolling time t_{NRT} is calculated together with the above failure time according to:

$$t_{NRT} = t_{GRT} - t_{RCT} - t_{FT}$$
$$t_{NRT} = 7{,}811 \, h \, a^{-1} - 533 \, h \, a^{-1} - 1{,}094 \, h \, a^{-1}.$$
(9.1.3)
$$t_{NRT} = 6{,}184 \, h \, a^{-1}$$

So, the yearly production of the HSM is:

$$P = 563 \, t \, h^{-1} \cdot 6{,}184 \, h \, a^{-1}$$
$$P = 3{,}482{,}829 \, t \, a^{-1}$$
(9.1.4)

Exercise 9-2: Energy demand for slab reheating in mill furnaces

A semi-continuously operating HSM with one reheating furnace produces 1.5 million tons hot strip per year. The hot charging rate of slabs in the reheating furnace is 20 %. The average hot charging temperature is 800 °C. Determine the cost benefit for the HSM due to hot charging of slabs considering the further boundary conditions below

- Specific heat capacity of slabs:
  $c_p = 365 + 0.327 \cdot T$; c_p [J kg^{-1} K^{-1}], T [K],
- Furnace discharging temperature:
  $T_{Charge} = 1{,}473.15 \, K = 1{,}200 \, °C$,

- Energy costs for slab reheating: C_{Energy}= 12 €/t for 100% cold charging rate.

Solution:

The specific energy for slab reheating in the furnace, **Fig. 9.1**, is given by

$$w = c_p \cdot (T - 273.15 \text{ K}). \quad (9.2.1)$$

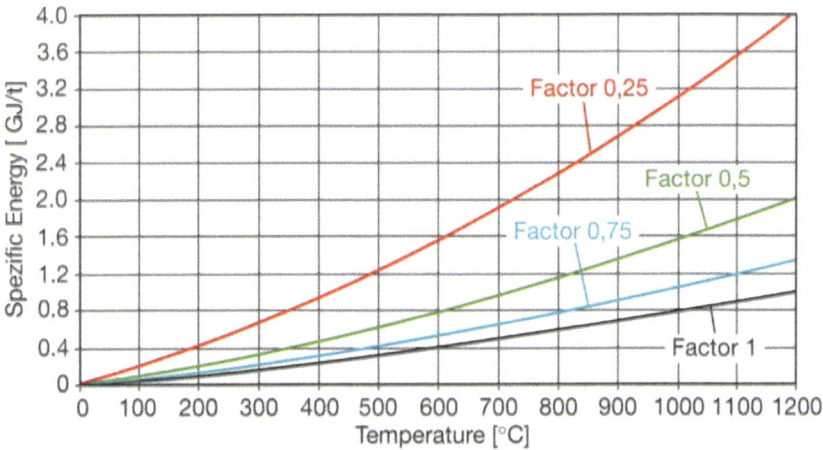

Fig. 9.1: Heating energy of carbon steel grade slabs in reheating furnaces. Parameter: Furnace efficiency factor

Inserting the values for hot charging and furnace discharging temperature in equation (9.2.1) gives:

$w_{hot} = w\ (1{,}073.15\text{ K} = 800\ °C) = 0.572\ \text{GJ t}^{-1}$ and
$w_{exit} = w\ (1{,}473.15\text{ K} = 1{,}200\ °C) = 1.016\ \text{GJ t}^{-1}$.

The difference of the specific reheating energies gives the reduction in case of 100% hot charging of slabs:

$$\Delta w = W_{exit} - W_{hot}$$
$$\Delta w = 0.444 \text{ GJ t}^{-1}$$
(9.2.2)

The related reduction of energy use in case of 100% hot charging therefore is:

$$\Delta w_{rel., 100\%} = \frac{W_{exit} - W_{hot}}{W_{exit}}$$
$$\Delta w_{rel., 100\%} = 1 - \frac{W_{hot}}{W_{exit}} = 43.7 \text{ \%}$$
(9.2.3)

For the part of 20% hot charging it is: $\Delta w_{rel., 20\%} = 8.74 \text{ \%}$.

The cost benefit for the given production of 1.5 million tons per year is 1.05 €/t or 1.57 million € per year.

Remark

In the following, practical and modeling results regarding reheating of slabs and billets in reheating furnaces of rolling mills are discussed. The types of treated reheating furnaces are walking beam, roller hearth and pusher type design. The nominal reheating capacity of the furnaces is in between 60 and 300 t/h.

In the data evaluation among others energy consumption and specific furnace hearth capacity are compared for different furnace types. The average energy use of the investigated furnaces is 1.45 GJ/t; this means, the furnace efficiency factor is roughly 70 %, Fig. 9.1.

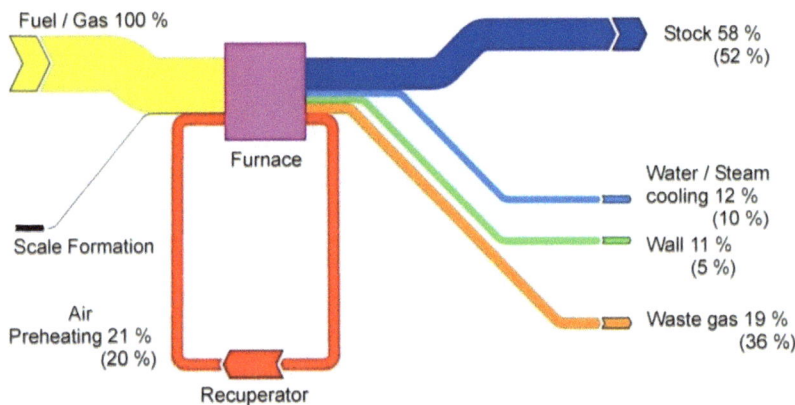

Fig. 9.2: Energy balance (with average values) in reheating furnaces. The numbers are values for bar and wire rod mills; numbers in parenthesis for HSM's /13/

The first step to find optimization potentials is the modeling of the energy and heat balance of the reheating furnaces. **Fig. 9.2** gives the average heat balance with data from evaluations for bar and wire rod mills and for HSM's. The numbers are related to fuel use: 58 % of the fuel is used for reheating the billet or slab, 12 % are losses caused by the furnace water and steam supplied cooling system, 11 % are losses at the furnace wall and 19 % waste gas losses. The differences found between the average numbers of both evaluations are minor than the differences found for the individual furnaces. E.g. the evaluated furnace cooling losses are between 7 % and 21 %, the furnace wall losses between 5 % and 19 %.

The following conclusions can be drawn from the investigations at site:

- The view of the contributions of individual heat losses in the heat balance equation gives the differences of the heat losses or the efficiency of heat recovery.
- Especially the different furnace operation modes have a significant influence on the heat balance. Rolling and production time, idle times with charged and empty furnace, idle times with furnace ready for discharging or heat conservation of the inserted stock have to be taken into account.

Fig. 9.3: Influencing parameters on specific energy use of mill reheating furnaces /13/

To get potential for improvement of furnace design and furnace operation practice the investigation of the influencing parameters by mathematical and physical models or statistics is performed on operational data. **Fig. 9.3** illustrates as a result of model research the influence of some parameters on the specific energy.

The specific energy of reheating furnaces is influenced by:

- Oxygen enrichment of the fuel air and improvement of air/fuel control,
- Increasing the temperature of in the furnace preheating zone,
- Improvement of furnace thermal isolation,
- Application of process models for furnace control,
- Lowering furnace discharging temperatures,
- Increasing furnace charging temperatures and
- Increasing furnace area hearth capacity.

10. Commissioning of rolling mills

General remarks regarding to this item are given in Chapter 17 of the book "Modern Hot Strip Production" /32/.

Exercise 10-1: Start-up curves and production

The start-up curve for production of mills is generally described by a non-linear function:

$$P(t) = a \cdot t^b \cdot \exp(c \cdot t) \qquad (10.1)$$

with

$$t_{startup} = -\frac{b}{c} \text{ (Duration of startup phase)}, \qquad (10.2)$$

$$t_{inflect} = -\frac{b - \sqrt{b}}{c} \text{ (Inflection point in time)}. \qquad (10.3)$$

From the derivative with respect to time of the function in equation (10.1) the time at which the production increase is a maximum can be calculated. The result for the respective time is given in equation (10.3).

a.) Calculate the parameters of the function for the scaling conditions $P(t_{startup}) = 1$ and $t_{startup} = 1$ and the two inflection points $t_{inflect} = 0.3$ and $t_{inflect} = 0.7$.

b.) Determine the resulting production loss of the plant in case of retarded start-up for two cases and assumptions to be compared:

(1) Linear start-up with the slope $m_1 = 1.2$ for $0 \leq t \leq 0.3$; Linear start-up with the slope $m_2 = 0.928$ for $0.3 < t \leq 1$.

(2) Linear start-up with the slope $m_3 = 0.761$ for $0 \leq t \leq 0.7$; Linear start-up with the slope $m_2 = 1.557$ for $0.7 < t \leq 1$.

c.) Calculate the tonnage and the loss in sales due to a retarded start-up for a hot skin pass mill with a nominal capacity of 60,000 t per month, a start-up time period of seven months and a sales price of 500 €/t.

Solution:

Exercise 10-1-a

Using equation (10.2) with the scaling condition $t_{startup} = 1$ it results

$$b = -c. \quad (10.4)$$

Inserting in (10.3) gives:

$$t_{inflect} = -\frac{b-\sqrt{b}}{c} = \frac{b-\sqrt{b}}{b} = 1 - \frac{\sqrt{b}}{b} = 1 - \frac{1}{\sqrt{b}}. \quad (10.5)$$

The coefficients can be determined by conversion:

$$b = \left(\frac{1}{1-t_{inflect}}\right)^2 \qquad (10.6)$$

and

$$c = -\left(\frac{1}{1-t_{inflect}}\right)^2. \qquad (10.7)$$

For the two inflection points follows:

$t_{inflect} = 0.3$: $b = 2.04$ and $c = -2.04$;

$t_{inflect} = 0.7$: $b = 11.1$ and $c = -11.1$.

For the calculation of the parameter a the second scaling condition $P(t_{startup}) = P(1) = 1$ is used:

(1) $t_{inflect} = 0.3$: $b = 2.04$ and $c = -2.04$;

$$\begin{aligned}P(1) &= 1 = a \cdot 1^{2.04} \cdot \exp(-2.04 \cdot 1) = a \cdot \exp(-2.04) \\ &\Leftrightarrow a = \exp(2.04) = 7.69\end{aligned} \qquad (10.8)$$

(2) $t_{inflect} = 0.7$: $b = 11.1$ and $c = -11.1$.

$$\begin{aligned}P(1) &= 1 = a \cdot 1^{11.1} \cdot \exp(-11.1 \cdot 1) = a \cdot \exp(-11.1) \\ &\Leftrightarrow a = \exp(11.1) = 66{,}171\end{aligned} \qquad (10.9)$$

The non-linear start up curves for the two inflection points can be finally written as:

$t_{inflect} = 0.3$:
$$P_{0.3}(t) = 7.69 \cdot t^{2.04} \cdot \exp(-2.04 \cdot t) \text{ for } 0 \leq t \leq 1 \quad (10.10)$$

and

$t_{inflect} = 0.7$:
$$P_{0.7}(t) = 66,171 \cdot t^{11.1} \cdot \exp(-11.1 \cdot t) \text{ for } 0 \leq t \leq 1. \quad (10.11)$$

Exercise 10-1-b

The total production $P_{tot_{0.3}}$ for the start up curve with the inflection point

$$t_{inflect} = 0.3 \quad (10.12)$$

Follows according to $P_{0.3}(0.3) = 0.36$ with

$$P_{1_{0.3}} = \frac{1}{2} \cdot (0.36 \cdot 0.3) = 0.054, \quad (10.13)$$
$$P_{2_{0.3}} = (0.36 \cdot 0.7) = 0.252 \text{ and} \quad (10.14)$$
$$P_{3_{0.3}} = \frac{1}{2} \cdot (0.64 \cdot 0.7) = 0.224 \quad (10.15)$$

to

$$P_{tot_{0.3}} = P_{1_{0.3}} + P_{2_{0.3}} + P_{3_{0.3}} = 0.53. \quad (10.16)$$

The total production $P_{tot_{0.7}}$ for the start up curve with the inflection point $t_{inflect} = 0.7$ follows according to

$P_{0.7}(0.7) = 0.53$ with

$$P_{1_{0.7}} = \frac{1}{2} \cdot (0.53 \cdot 0.7) = 0.1855, \quad (10.17)$$

$$P_{2_{0.7}} = (0.53 \cdot 0.3) = 0.159 \text{ and} \quad (10.18)$$

$$P_{3_{0.7}} = \frac{1}{2} \cdot (0.47 \cdot 0.3) = 0.0705 \quad (10.19)$$

to

$$P_{tot_{0.7}} = P_{1_{0.7}} + P_{2_{0.7}} + P_{3_{0.7}} = 0.415. \quad (10.20)$$

The production loss is:

$$\Delta P = P_{tot_{0.3}} - P_{tot_{0.7}} = 0.115 = 11.5\,\%. \quad (10.21)$$

Exercise 10-1-c

For the hot skin pass mill with a monthly production capacity of 60,000 t and a start- up phase of seven months results an absolute production loss of (7 months x 6,900 t/month) ≈ 50,000 t. With the sales price of 500 €/t this makes a sales loss of approximately 25 million Euros.

11. Materials for rolls and wear of rolls

Exercise 11-1: Roll consumption and roll service life time

In a HSM HCr (high chromium) work rolls are used at the FS's F1 to F4. The maximum roll diameter is 820 mm and the minimum diameter 720 mm. The average rolled tonnage before roll scrapping is 350,000 t. The ineffective roll wear is 500 t/mm. Determine the numbers:

a) The grindable roll shell thickness,

b) The effective roll wear,

c) The theoretical rolled tonnage and

d) The number of rolling schedules with an average rolled tonnage of 3,000 t per rolling campaign.

Solution:

Exercise 11-1-a

The grindable roll shell thickness is given by the half of the difference of the maximum and minimum roll diameter. Using the given data it is 50 mm.

Exercise 11-1- b

The effective roll wear can be calculated by the ratio of the average tonnage until roll scrapping (350,000 t) and the grindable roll shell thickness (50 mm). The value is 7,000 t/mm.

Remark

The effective roll wear is measured in the roll shop and contains the values due to pure roll wear (theoretical roll wear) and the necessary roll grind due to damages (ineffective roll wear). Roll damages can be caused by roll material failures or by the rolling process, e.g. roll marks by shearing/doubling of the strip. During rolling roll damages can occur because of insufficient roll cooling or failures of the corresponding roll cooling system. As guideline for the roll temperature a temperature range between 60 °C and 80 °C (measured after finishing the rolling campaign) depending on the roll grade material is recommended. The value of the roll temperature can be measured manually using a thermocouple. For each roll a sheet for documentation exists in the roll shop. In this sheet the complete life cycle of the roll is noted including the wear data and corresponding reasons.

Exercise 11-1-c

The theoretical roll tonnage is given by multiplication of the theoretical roll wear and grindable roll shell thickness. Theoretical roll wear is 7,500 t/mm according to the exercise data and will be calculated as the sum of effective and ineffective roll wear. With the roll shell thickness of 50 mm the corresponding tonnage is 375,000 t.

Exercise 11-1-d

The number of rolling schedules is given by the ratio of the average tonnage until scrapping of the roll (375,000 t) and the average tonnage per rolling schedule (3,000 t). The number of rolling campaigns until scrapping is n = 125.

12. Questionnaire of rolling mill technique

Exercise 12-1: Terms and definitions

a.) What is the name of the combined casting and in-line rolling technology?

☐ 1	CSP-technology
☐ 2	BVB-technology

b.) What is the name of the technique of width map-based pilot control and width control in HSM's?

☐ 1	AGC-technique
☐ 2	AWC-technique
☐ 3	PVC-technique
☐ 4	SSC-technique

c.) What is the name of a HSM with five one way operating RMS's and seven FMS's?

☐ 1	CSP-mill
☐ 2	Fully continuous operating-HSM
☐ 3	3/4-HSM
☐ 4	Semi-continuous-HSM

d.) Which actuators and techniques exist in HSM's for the adjustment and control of the thickness profile and flatness?

☐ 1	CVC-technique
☐ 2	Roll bending
☐ 3	HGC-technique

e.) By which factor is the distance of a descaler to the strip surface to be changed in order to double the impact of the descaler? (Descaling pressure, flow rate and type of descaling nozzle remain constant)?

☐ 1	$\sqrt{2}$
☐ 2	2
☐ 3	$\dfrac{1}{\sqrt{2}}$
☐ 4	$\dfrac{1}{2}$

f.) What is meant by „step-control"?

☐ 1	Controlled lifting of the wrapper rolls at the DC during coiling of the strip head end
☐ 2	Spreading of the coiler mandrel during strip coiling phase
☐ 3	Acceleration of the main drives in the FM for compensation of the temperature wedge over the strip length
☐ 4	Rolling using CB operation

g.) Which measuring techniques for registering material width are common in HSM's?

☐ 1	Laser triangulation methods
☐ 2	Stereoscopic techniques
☐ 3	Irradiation methods based on X-ray- or isotopic radiation beams
☐ 4	Techniques based on pyrometers
☐ 5	Radar technique

h.) What is meant by „skid-mark"?

☐ 1	Cooled skids for slab support in reheating furnace
☐ 2	Local temperature decrease in the slab or strip caused by cooled skids in the reheating furnace
☐ 3	Local surface defects onto the slab caused by slab transport in a pusher type furnace
☐ 4	Local formation of scale defects on the final strip caused by clogged descaling nozzles

i.) Which correction functions are taken into account in the "Gagemeter equation"?

☐ 1	Roll thermal expansion
☐ 2	Roll wear
☐ 3	Water appearance in the thickness measuring area
☐ 4	Roll bearing oil film
☐ 5	Roll eccentricity
☐ 6	Work roll bending
☐ 7	Strip temperature variations due to skid-marks
☐ 8	CVC-roll shifting position (equivalent mechanical work roll grind)

j.) Which rolling products belong to the group of "Flat products"?

☐ 1	Hot strip
☐ 2	Cold strip
☐ 3	Steel bar products
☐ 4	Wire rod products
☐ 5	Light and heavy sections
☐ 6	Heavy plates
☐ 7	Rings

k.) Which dimension is equal to the unit 1 MPa?

☐ 1	1 N mm^{-2}
☐ 2	1 kN mm^{-2}
☐ 3	1 N m^{-2}
☐ 4	1 MN mm^{-2}
☐ 5	1 kN m^{-2}

l.) Which plant concept is common applied nowadays in the world for new CRM's?

☐ 1	Fully continuous operating Tandem Mills (TCM)
☐ 2	Coupled Pickling and Tandem Rolling Mills (PLTCM)
☐ 3	Coupled Pickling-/Tandem-/Annealing Lines

m.) Which performance is possible when rolling cold strip on a CCM (Compact Cold Mill) with two stands?

☐ 1	The possible yearly production capacity is roughly 0.8 million tons
☐ 2	The possible yearly production capacity is roughly 0.4 million tons
☐ 3	The possible yearly production capacity is roughly 1.2 million tons

n.) What is the name of the s-shaped and shiftable rolls implemented in a CRM stand?

☐ 1	PVC-rolls
☐ 2	BVB-rolls
☐ 3	EDC-rolls
☐ 4	CVC-rolls
☐ 5	CVGL-rolls

o.) Which kind of feed stock is used for HDGLs (Hot dip galvanizing lines) in cold rolling plants?

☐ 1	Non-pickled hot wide strip
☐ 2	Pickled hot wide strip
☐ 3	Annealed and skin passed cold strip
☐ 4	Cold rolled strip

p.) The unit 1-I-Unitcharacterizes a flatness defect of hot and cold strip and means are fiber length difference of

☐ 1	100 µm/m
☐ 2	10 µm/m
☐ 3	1 µm/m

q.) Specify complete orthogonal systems for mathematical description of measured flatness defects for flatness control in CRM's?

☐ 1	Polynomials of n-thorder
☐ 2	Legendre polynomials
☐ 3	Tschebycheff polynomials

r.) What kind of thickness defects are transferred when applying thickness control (AGC) in HSM's in case of an ideal infinite non-rigid mill stand (Mill stand modulus $C_G \to 0$)?

☐ 1	Eccentricities of back up and work rolls
☐ 2	Variations of the yield stress caused by differences of the chemical composition in the cold strip
☐ 3	Variations of the loaded roll gap caused by thermal work roll expansion
☐ 4	Thickness variations due to thickness defects of the entry strip
☐ 5	Variations of the loaded roll gap caused by roll wear

s.) Which principles and methods are common to provide load safety during coil transport?

☐ 1	Force closure during packaging
☐ 2	Form closure during packaging
☐ 3	The combination of force and form closure during packaging
☐ 4	Coil connection via the respective coil eyes with a rope and fastening the rope on the truck tailboard
☐ 5	Usage of coil-bodies and anti-slipping mats

t.) At which aggregate in a cold rolling plant the final mechanical-technological properties are adjusted?

☐ 1	Pickling Line
☐ 2	Tandem Mill
☐ 3	Hot dip galvanizing line with an inline skin pass mill
☐ 4	Skin pass mill after batch annealing facility
☐ 5	Cut-to-length line in a finishing line

Solution:

Exercise	Solution
a	☑ 1
b	☑ 2, ☑ 4
c	☑ 2
d	☑ 1, ☑ 3
e	☑ 3
f	☑ 1
g	☑ 2, ☑ 5
h	☑ 2
i	☑ 1, ☑ 2, ☑ 4, ☑ 5, ☑ 6, ☑ 7, ☑ 8
j	☑ 1, ☑ 2, ☑ 6
k	☑ 4
l	☑ 1, ☑ 2
m	☑ 1
n	☑ 4
o	☑ 1, ☑ 2
p	☑ 2
q	☑ 1, ☑ 2
r	☑ 2, ☑ 4
s	☑ 1, ☑ 2, ☑ 3, ☑ 5
t	☑ 3, ☑ 4

13. Bibliography

This chapter gives a list of the recommended literature regarding the topic „Technology of Hot and Cold Strip Rolling".

/1/: Kopp, R., Wiegels, H.: Einführung in die Umformtechnik. 1. Auflage Aachen, Verlag der Augustinus Buchhandlung, Pontstraße 96, 52062 Aachen, 1998.

/2/: Autorenkollektiv: Rationeller Energieeinsatz bei Umformprozessen. VEB Deutscher Verlag für Grundstoffindustrie, Leipzig 1981.

/3/: Hensel, A.; Spittel, T.: Kraft- und Arbeitsbedarf bildsamer Formgebungsverfahren. VEB Deutscher Verlag für Grundstoffindustrie, Leipzig 1978.

/4/: Lange, K.: Umformtechnik. Springer-Verlag, Berlin.
Band 1 – Grundlagen (1984).
Band 2 – Massivumformung (1988).
Band 3 – Blechbearbeitung (1990).

/5/: Lippmann, H.; Mahrenholtz, O.: Plastomechanik der Umformung metallischer Werkstoffe. Springer Verlag, Berlin 1967.

/6/: Spur, G.; Stöferle, T.:
Handbuch der Fertigungstechnik, Band 2 (1983-1985). Carl Hanser Verlag, München

/7/: Verein Deutscher Eisenhüttenleute: Grundlagen der Bildsamen Formgebung. Verlag Stahleisen 1966.

/8/: Verein Deutscher Eisenhüttenleute: Werkstoffkunde Stahl. Verlag Stahleisen 1984.

/9/: Weber, K. H.: Grundlagen des Bandwalzens. VEB Deutscher Verlag für Grundstoffindustrie, Leipzig 1974.

/10/: Hoff, H.; Dahl, T.: Grundlagen des Walzverfahrens. Verlag Stahleisen 1955.

/11/: Wusatowski, Z.: Grundlagen des Walzens. VEB Deutscher Verlag für Grundstoffindustrie, Leipzig 1963.

/12/: Pawelski, O.: Theoretische Grundlagen des Warmwalzens. Mitteilung aus dem Max-Planck Institut für Eisenforschung, Abhandlung 1174 in: Herstellung von Halbzeug und warmgewalzten Flacherzeugnissen, Verlag Stahleisen, Düsseldorf.

/13/: Degner,M.; Mauk, P.J., u.a.: Spitzentechnologien im Wettbewerb bei der Erzeugung von Flach- und Langprodukten. Stahl und Eisen 3/08.

/14/: Degner, M.: Mathematik für Physiker und Ingenieure – Grundlagen. Buchveröffentlichung. Verlag Stahl Eisen, April 2008.

/15/: Degner, M.: Technologie der Warmbanderzeugung. Verlag Stahl Eisen, September 2008.

/16/: Degner, M.: Untersuchungen zur Optimierung der Maßhaltigkeit von warmgewalzten Bändern. Dr.-Ing. Dissertation TU-Clausthal, April 1993.

/17/: Adlung, H.; Degner, M.; Luttmann, R.; Neuschütz, E.; Tamler, H.: Walzen von gleichmäßigem Dickenprofil. EGKS-Abschlußbericht 1995, Unterausschuss D2 „Walzen von Flachprodukten".

/18/: Degner, M.; Luttmann, R.; Palkowski, H.; Thiemann G.: Untersuchungen zum Einfluss des Laageraufschwimmens auf den Lastwalzspalt. Stahl und Eisen 3/96.

/19/: Degner, M.; Müller, U.; Thiemann, G.; Winter, D.: Topometric on-line flatness measuring system for hot strip. Metallurgical Plant and Technology (MPT) 6/98.

/20/: Degner, M.; Thiemann, G.; u. a.: Erfassung des Coilstirnprofils von warmgewalzten Bändern. Stahl und Eisen 10/99.

/21/: Degner, M.; Thiemann, G.; u. a.: Measurement of the transverse temperature profile of hot rolled strips. Metallurgical Plant and Technology (MPT) 2/00.

/22/: Degner, M.: Advanced measuring techniques for production of ultra thin hot strip. Proceedings Symposium „Hot Rolling of Thin-Gauge Sheet Steel" der „Iron and Steel Society (ISS)", Mai 2000 in Toronto, Ontario, Kanada

/23/: Degner, M.; Thiemann, G.; u. a.: Use of optimized descaling nozzles on the wide HSM of Thyssen Krupp Stahl. Metallurgical Plant and Technology (MPT) 5/00.

/24/: Degner, M.; Thiemann, G.: Developments in production of ultra thin gauge hot strip. Metallurgical Plant and Technology (MPT) 1/01.

/25/: Palkowski, H.; Degner, M.: Schmierung im Walzspalt bei der Fertigung von Warmbreitband. Der Kalibreur, Heft 63, Juni 2002.

/26/: Degner, M.; Thiemann, G.: Entwicklung der Coilboxtechnologie am Beispiel der Warmbreitbandstraße in Bochum. Stahl und Eisen 10/02.

/27/: Degner, M.; Palkowski, H.: Basics in Flat Rolling, 2016. IMP InterMediaPartners GmbH, Wuppertal.

/28/: Degner, M.; u. a.: Skimming through 5 generations of hot rolling. Steel Grips No. 3, May/June 2003.

/29/: Degner, M.; u.a.: Umformtechnik. In Chemische Technik – Prozesse und Produkte. Winnacker Küchler 5. Auflage, Band 6a Metalle. Wiley-VCH Verlag, 2006

/30/: Degner, M.; u.a.: Entwicklungen und Tendenzen in der Stahlindustrie. Der Kalibreur, Heft 67, Juni 2006.

/31/: Degner, M.: Produktionsentwicklung von warm- und kaltgewalzten Produkten und Analyse der Anlagenkapazitäten. Jahrbuch Stahl 2007. Herausgeber Stahlinstitut VDEh, Verlag Stahl Eisen.

/32/: Degner, M.: Modern Hot Strip Production. Verlag Stahleisen, Düsseldorf, November 2013.

/33/: Autorenkollektiv: Steel Manual. Verlag Stahleisen, Düsseldorf 2007.

/34/: Spur, G.: Handbuch Umformen, 2012. Carl Hanser Verlag, München

14. List of key words

3/4-continuous HSM ... 22, 31
absolute measuring error 83
absolute thickness draft 20, 48, 73, 107
AGC 69, 125, 131
attenuation law 93, 96
AWC 125
backup roll eccentricity 74
bar and wire rod mills. 114
billet 114
biting angle 47
biting condition 41, 47
bottle neck 25
CCD-camera 83
coilbox (CB) 7, 25, 26, 39, 41, 43, 44, 45, 126
cold charging rate 112
cold rolling mill(CRM) 6, 7, 8, 15, 16, 19, 20, 28, 106, 129, 130
Compact Cold Mill (CCM) 129
compression test 67
crop shear (CS) 7, 33, 34, 35, 36, 61, 62, 63
CSP-technology 125
CVC 125
deformation energy 52
deformation rates 67
deformation resistance . 69
deformation speed 49
deformation stress 52
density .. 12, 14, 39, 58, 60
discharging temperature 53, 112, 116
down coiler (DC) 7, 33, 34, 35, 36, 39, 40, 126
effective roll wear 123
effective strain 59, 60
error propagation 72
exposed automotive applications 28
failure time 110
final rolling speed 8
finishing mill (FM) .. 7, 8, 9, 10, 12, 25, 26, 32, 33, 34, 35, 36, 39, 48, 63, 72, 83, 86, 125, 126

finishing stand (FS) . 7, 13, 16, 18, 25, 33, 34, 35, 36, 37, 40, 48, 72, 122
flat products 128
flattened roll radius 20
flying strip 85, 90, 102
fully continuous HSM 31
furnace descaler 35
furnace discharging temperature 111

furnace preheating zone 116
furnace thermal isolation 116
gagemeter equation 75
geometry factor 49
gravity force 37
gross production time ... 30
gross rolling time 110, 111
gross time factor 31
heat balance 51
heat conduction 52, 56, 58
heat radiation ... 52, 54, 59
heat transfer coefficient 57
heavy plates 128
high chromium work rolls 122
Hitchcock 105, 107
hot charging 111, 113
hot dip galvanzing line 130
hot flow curve 64
Hot plate mill (HPM). 7, 13
Hot strip mill (HSM) .. 6, 7, 8, 11, 13, 14, 15, 16, 18, 22, 23, 25, 31, 33, 34, 35, 36, 37, 39, 42, 47, 48, 50, 51, 69, 83, 86, 87, 106, 110, 111, 114, 122, 125, 127, 136
hot compression test 64
hysteresis 69, 70, 71
idle time 110, 115
idle times 110
impact 126

ineffective roll wear 122
intercept theorem 83
I-Unit 130
laser triangulation 127
law of volume constancy 55
Legendre polynomials 130
lever arm method 50
light and heavy sections 128
load safety 131
main strain 59, 60
material modulus ... 69, 70, 73, 76, 79
measuring error 83
mechanical-technological properties 132
mill stand equation 75
mill stand modulus 70
mill stand stretch 75, 86
mill-pacing 26
Multi Roller Stand (MRS) 7, 108
naked eye 27
net rolling time 110
non-productive time 30
O5-grades 28, 29
O5-strip material 28
pass schedule . 25, 26, 31, 32, 34, 48
Pawelski 57, 134
PLTCM 129
radiation area ... 43, 44, 45
radiation loss 55

relative measuring error 83
relative thickness draft 47, 49
relative thickness reduction 20, 35
rigid mill stand 80
rings 128
roll bearing oil film 128
roll bite length 49, 51
roll consumption 122
roll eccentricity 69, 128, 131
roll flattening 105, 107
roll force 19, 20, 48, 49, 81
roll gap geometry 47
roll gap length 56
roll peripheral speed 48
roll service life time 122
roll thermal expansion. 69, 128
roll torque 19, 20, 50
roll wear 128
roughing mill (RM) 7, 12, 23, 25, 26, 32, 34, 51, 61, 63, 86, 125
roughing stand (RS) 7, 13, 25, 31, 32, 33, 34, 35, 36, 39, 51, 61, 86, 87
s-defect 78
semi continuous HSM .. 31
Sims 48
Sims roll gap model 48
skid-mark 82, 127

skin pass mill 28, 118, 121, 132
slab 114
sliding friction 107
slip forward 48, 52
specific heat capacity .. 53, 60, 61, 62, 111
specific reheating energy 112
speed-up 39, 41, 42, 63
spezific roll force 105
SSC 125
start-up curve 117
step-control 126
stereoscopic measuring system 86
strip tracking 86
technology of casting and rolling 125
temperature compensation 96, 98
thermal conductivity 59
thickness control 72, 75
thickness draft 75
thickness error 69, 72
thickness profile wedge 86
thickness reduction 47
transfer bar 8, 9, 10, 26, 32, 43, 44, 45, 46, 47, 72, 86
transfer factor 73
Trinks 50
Tschebycheff polynomials 130

utilization ratio 31
von Mises 59
water steam 94
width measurement 83
wire rod 128
work roll bending 128

wustite 92
yield strength ... 20, 48, 49, 74, 82
yield stress 20, 86, 107, 131